Praise for

The Farm Then [illegible]

Douglas Stevenson has shown us how a com[illegible]ity of 1960s counte[illegible]tural idealists could come to grips with the outside world but still k[illegible] much of their original dream intact, and provides us with an absorbing overview of this spiritual community and ecovillage that continues to embody the best values of alternative America. It is a compelling story, very well told.

—Timothy Miller, Professor of Religious Studies, University of Kansas and author, *The 60s Communes*

This is a fascinating inside look at an ecovillage which has endured for decades, maturing from its initial foundation and ad hoc arrangements to become a skilled group of people living in tune with nature and forming an integral part of the rural economy of the area. Douglas Stevenson clearly lays out the challenges faced by The Farm and how these were met over many years. This is essential information needed to get new ecovillages off on the right foot from the beginning.

—Nicole Foss, Senior Editor, *The Automatic Earth*

Doug Stevenson has written by far the best history and interpretation of The Farm that I know of. Organized topically, it deals with every major aspect of this important intentional community. He shows these both in their historic development and their concrete embodiments. Told from an insider's view, the book is candid and deeply informed.

—Robert J Rosenthal, PhD, Professor/Chair of Philosophy, Hanover College, Hanover IN

The Farm is a fascinating account of one of the most important and long-lasting experiments in intentional community. Honest, extensive and informative, it's a great read and should be on the must-read shelf for anyone interested in community, group dynamics and the history of social movements.

—Starhawk, activist, permaculture designer and teacher and author, *The Fifth Sacred Thing* and *The Empowerment Manual*

Members of the Farm doggedly reconciled spiritual ideals and secular realities, weathered the upheaval of their governance and economy, stayed loyal to their land, and endured unto the third generation, with the promise of more to come. It's among the most inspiring social experiments of our time, a righteous great hippie accomplishment.

In clear unstudied prose, Douglas Stevenson has done great service, distilling and reporting his intentional community's decades of learning, trials, errors and myriad details of their tangible achievements. Both a manual and an origin saga, *The Farm Then and Now* is a fascinating, highly consequential book.

—Stephanie Mills, author, *Epicurean Simplicity* and *On Gandhi's Path: Bob Swann's Work for Peace and Community Economics*

Douglas Stevenson has woven together an insider's look at The Farm, holding nothing back. For anyone who may have fantasized what might have happened if they had just gone off and joined a hippie commune, here it is: the good, the bad, and the sublimely naive.

—Albert Bates, author, *The Post-Petroleum Survival Guide and Cookbook* and *The Biochar Solution.*

DOUGLAS STEVENSON

THE FARM THEN AND NOW

A MODEL FOR *Sustainable Living*

Cover design by Diane McIntosh.
For full listing of cover and interior image credits, see page 228.

Printed in Canada. First printing March 2014.

New Society Publishers acknowledges the financial support of the Government of Canada through the Canada Book Fund (CBF) for our publishing activities.

Inquiries regarding requests to reprint all or part of *The Farm Then and Now* should be addressed to New Society Publishers at the address below.

To order directly from the publishers, please call toll-free (North America) 1-800-567-6772, or order online at www.newsociety.com

Any other inquiries can be directed by mail to:

New Society Publishers
P.O. Box 189, Gabriola Island, BC V0R 1X0, Canada
(250) 247-9737

LIBRARY AND ARCHIVES CANADA CATALOGUING IN PUBLICATION

Stevenson, Douglas, 1953–, author
The Farm then and now : a model for sustainable living / Douglas Stevenson.

Includes index.
Issued in print and electronic formats.
ISBN 978-0-86571-769-5 (pbk.).—ISBN 978-1-55092-565-4 (ebook)

1. Farm (Summertown, Tenn.). 2. Farm (Summertown, Tenn.)—History. 3. Sustainable living—Tennessee—Summertown—Case studies. 4. Sustainable agriculture—Tennessee—Summertown—Case studies. 5. Cooperative societies—Tennessee—Summertown—Case studies. 6. Counterculture—Tennessee—Summertown—Case studies. 7. Summertown (Tenn.)—Social life and customs—Case studies. I. Title.

GE198.T45S74 2014 333.7209768'432 C2013-908563-7
C2013-908564-5

New Society Publishers' mission is to publish books that contribute in fundamental ways to building an ecologically sustainable and just society, and to do so with the least possible impact on the environment, in a manner that models this vision. We are committed to doing this not just through education, but through action. The interior pages of our bound books are printed on Forest Stewardship Council®-registered acid-free paper that is **100% post-consumer recycled** (100% old growth forest-free), processed chlorine-free, and printed with vegetable-based, low-VOC inks, with covers produced using FSC®-registered stock. New Society also works to reduce its carbon footprint, and purchases carbon offsets based on an annual audit to ensure a carbon neutral footprint. For further information, or to browse our full list of books and purchase securely, visit our website at: www.newsociety.com.

Contents

Introduction

In late 1960s San Francisco, a former university professor turned hippie guru set in motion a chain of events that would lead to one of the most dynamic social experiments of modern time. Like so many of his contemporaries, Stephen Gaskin was a catalyst for something much larger than himself, one component in a cultural upheaval with impact on the world at large, the founder of an iconic symbol representing the belief in a higher purpose, where people come together for the greater good, to launch a new society where peace and cooperation are the status quo.

Stephen Gaskin's Monday Night Class was a gathering of the psychedelic mind, a random collection of the best and brightest of a generation, in a quest for knowledge and understanding. From this number, a core group of true seekers set forth on an epic journey across the country in 60 school buses, a Caravan on a mission of peace and love.

Tennessee became the promised land, a place to put ideas and ideals into practice. It was here on 1,700 acres of forest and fields that babies would be born, crops grown and a town built from nothing but a collective dream and a lot of sweat.

The Farm had one purported goal: change the world.... And in many ways it did. But the world also changed The Farm, and in order to survive, it had to adapt.

The Farm Community was founded in 1971 with the purest of intentions—that all who came would be cared for, fed, clothed, healed, provided shelter—referencing the Book of Acts so that its neighbors could grasp what it was these hippie kids were trying to do.

> *All that believed were together, and had all things in common; And sold their possessions and goods, and parted them to all men, as every man had need.* (Acts 2:44-45)

There was a delicious euphoria, an energy so alive in the birth of something new, the momentum of a thousand strong behind a shared vision. The revolution was happening, and it was a blast! A rock and roll tribe, on a quest for enlightenment and planetary consciousness, in service to the world.

But by the fall of 1983, the dream had lost its luster. Deep in debt, its members disillusioned, frustrated and no longer willing to endure a self-imposed vow of poverty, The Farm made a radical shift that will be forever known as The Changeover.

Support from the community was over. Everyone was left on their own. Literally hundreds ran as fast and far away as they could.

A core group remained on the land, saving it from foreclosure. By the early '90s, the community was debt-free and back on its feet. The century was turning, and anything was possible. The future lies ahead.

The Farm's survival for over 40 years is a testament to patience and perseverance, what is possible when remaining true to your ideals in the face of endless obstacles. The community is a new-age hybrid, a blend of rural and high-tech lifestyles, classic individualism and the power of collectivity. It remains a flagship, a model of how we as planetary citizens may choose to live, with lessons to be learned from its successes and its failures, its weaknesses and its strengths. The Farm is an ongoing experiment on how human beings can be together in a meaningful and personal way, connected to the natural world.

The Farm is not in an isolated bubble, a glass dome. It is tied to and part of the larger community outside its borders, both local and state, country, the greater society and planet on which it exists. It is affected by world politics, the economy, weather, modern trends and ancient traditions, a microcosm of the big picture.

The word "community" has become a buzzword and in the process can lose its deeper meaning. Any collection of people gathered together around an element in common, be they online, in a chat room or sharing a game, is labeled as community. These identities can fill a void that is no longer satisfied by the actual place where people live, the isolation generated by the confines of urban life, the breakdown of the family, fractured and scattered, an excessive emphasis on individualism and the shallowness present in mainstream culture. It leaves people hungry for something more.

The phrase "intentional community" represents small groups of people who choose to live together in one place and share more aspects of their lives in a direct and tangible way. You are there not because you liked the house or the school was near where you wanted to live. You are making a conscious decision to share your life with more people.

Living in community touches every aspect of how we as humans relate with each other: How decisions are made and followed through, the essence of government. How we care for each other from beginning to end. How we support ourselves and earn a living, the homes in which we live. How new members are brought into the community, and about learning how to get along.

Stephen Gaskin once said, "The revolution is not about taking over the government, but taking over the government's function. We seceded as far as we could without them sending in the pony soldiers."

Creating community is about creating and developing workable, functioning alternative systems that restore our sense of purpose, empower the family and bring us closer to the natural order of things. On 1,700 acres in Middle Tennessee, a small group of people have tried to do just that.

Hey Beatnik, This Is The Farm Book was published in 1974 to illustrate what was happening on this land inhabited by a bunch of hippie idealists. It described the many different parts of community, how they all fit together, how each worked and why. In a very real sense,

Hey Beatnik became a blueprint for intentional community, a handbook for getting started.

The Farm Then and Now picks up the conversation some 40 years later. It reexamines the building blocks of community and their evolution through The Farm's history, and more importantly identifies how they function in the context of the community today.

You should know that, in writing this book, I am not on the outside looking in. I have been part of The Farm for over 40 years and am proud of what it has accomplished and what the community is today. But this book is not about me being a cheerleader. Any examination of this ongoing experiment in how people live together has to include the struggles, the mistakes, the problems and lessons to be learned that arise in every community.

Most of all, I hope this book inspires you to take a look at where you are, where you're going and where you want to be. May you be brave, and move forward, taking the next step that will get you there.

CHAPTER 1

Governance

FOR ANY ORGANIZED GROUP of people, the process of decision making is at the core of how it functions and moves forward. Some might say the challenge is even greater when one of their uniting principles is to question authority. Over the course of its history, The Farm has exemplified a range of systems, often simultaneously. The Farm's greatest success, the one that has ensured its survival through multiple decades, was its ability to move beyond a central charismatic leader into a working system where each person has the opportunity and free will to control not only their own destiny, but also the direction of the community as it moves toward the future.

Teacher and Class

Going back to the very beginning, the original model around which people gathered was built on the familiar relationship of teacher and classroom. Stephen Gaskin left the world of formal academia but used this structure to begin a group analysis of consciousness and human relationships, when, in 1968 he began hosting discussions as part of San Francisco State's Free University. As the after-hours class grew from a handful of people to over a thousand, gathering every Monday night, a need arose to establish ground rules. They were simple. Each person was given the opportunity to speak without interruption.

Common courtesy. Wait your turn. However, it was clear that Stephen was both moderating and leading the conversation, which consisted primarily of questions to and answers from him.

During this period in San Francisco, Stephen took certain steps to blur the line of teacher and student, while using the skills and experience he had acquired as a university professor to maintain order and direction. Rather than lecture from a podium on a stage, at Monday Night Class, Stephen sat on a lower-level platform, just slightly above the audience. As the primary focus of attention for those assembled, he channeled the energy of the crowd and distilled its essence, interpreted lessons learned and articulated these back to the assembled group, identifying when this new awareness resonated with greater principles.

Stephen Gaskin, founder of The Farm Community

In Eastern philosophy, the role or definition of "teacher" went beyond formal academic education dealing with the material world to include the role of spiritual instructor. Spiritual knowledge attained through centuries of study and application could be learned, taught and passed down from one generation to the next in the form of principles that serve as essential guideposts to be utilized throughout the course of a lifetime.

During the last century, individuals serving as representatives from various philosophies made their way from East to West, rising to a peak in the 1960s and '70s. Disillusioned with the hypocrisies and limitations of institutionalized Christianity, the youth of this period sought new answers and guidance, open to the solutions being put forward by these spiritual teachers and teachings emanating from India, Tibet, Japan and other Eastern countries, as well as from Native Americans and other Indigenous cultures.

And so it was that a range of self-proclaimed or officially designated spiritual teachers were working in the US and Europe. In a sense, all of them were competing to amass followings and establish themselves as teachers who could provide answers and a clear direction on the path to greater understanding of each individual's role and relationship to the universe.

These defined roles were brought into focus at an event in Boulder, Colorado, in the summer of 1969, called the Holy Man Jam. Yogi Bhajan, a Sikh from India, was a spiritual leader and entrepreneur who introduced Kundalini Yoga to the United States. Swami Satchidananda was widely known after his appearance at the Woodstock Music Festival in 1969. Representatives of Japanese Buddhism and a number of other teachers used this platform to attract a following. In a very real way, his high-profile appearance transformed Stephen from a counterculture philosopher to spiritual teacher, a person that would formally receive students committed to accepting him as a mentor. He distinguished himself from the others by acknowledging the truths to be found in all religions and spiritual philosophies, blending these into a new universal set of teachings that would relate directly to modern life and Western culture.

Warren (last names have been omitted to preserve the anonymity of individuals), a participant in the group meetings in San Francisco, remembers:

> Monday Night Class and our own inner experiences provided some answers but also generated more questions. Stephen was really in a very similar boat but older. He was really good at public speaking, crystallizing the issues of the time and channeling the energy of the group. I think both he and we misunderstood this. It looked like he had the "answer," when really what he had was the energy of the group. He was a psychedelic father figure who basically said, "Come be a part of this new family where you'll be understood and accepted and you'll be given the opportunity to grow spiritually."

Defining Roles on The Caravan

The roles of leadership and organization became more defined as Stephen and his "students" left San Francisco in the spring of 1970 to embark on the cross-country speaking tour that has become known as The Caravan. What had begun as a weekly meeting, connected by an extensive network of personal relationships, was taken to a new level through daily interaction and increased responsibility. It was no simple task to coordinate the movement of a couple of hundred people travelling in an array of school buses, delivery vans, cars and trucks on a daily basis. Functioning as a village on the move, the needs of humans to be fed and cared for had to be met each and every day, a very real challenge for youth only beginning to learn how to take care of themselves.

As the public face of the travelling band of new-age gypsies, Stephen took the first steps in establishing standards of behavior and structure. For example, to maintain order and a cohesive appearance, The Caravan needed to depart as a unified group. This meant that each morning it was important that everyone begin the activities together and be ready to leave for the next destination at the same time. Stephen explains:

> I started going around in the morning with a steel wrench tapping on the bumper of each bus, letting folks know that it was time to get up and moving. After a few days, one of our guys came up and said, "I can do that," and took over that task. And that's how it went. People were watching me to learn what needed to be done and figure out how they could step up to the plate and take on some of the responsibility for The Caravan.

While Stephen was recognized as a central authority figure, The Caravan was kept moving through the broader effort of people working together to achieve a common goal. Over the next year, more and more individuals stepped into positions of responsibility in order to keep the mobile village alive and functioning. Work crews had to be organized to generate money for gas and food. Anyone with mechanical skills became invaluable and immediately transformed into a teacher of a unique mystical order, the mechanic. It became clear that the same focused attention and discipline that defined the spiritual path had to be applied in order to keep engines starting and buses rolling.

Living on the Land—The Straw Boss

In the spring of 1971, when The Caravan came to rest in Tennessee, the number of tasks and roles to be filled multiplied a hundredfold. Although Stephen was generally regarded as the primary person in charge, there was plenty of room for others to exert their influence and establish positions of power within the informal hierarchy that was developing.

There was an immediate need to empower decision makers and surround them with people who would follow their direction, working together to implement the tasks at hand, as well as move toward the broader vision of self-sufficiency and social change. Along the way, The Farm adopted terms to define these persons so that their positions would be clear and not constantly challenged and up for debate. Crew chiefs were called "straw bosses," with each one representing a different aspect of community development or an important function. Larger groups, such as those dedicated to farming and raising food, might have several straw bosses, each one working with a crew of four or five to take on a specific role or manage a particular crop.

The various crew chiefs and straw bosses would meet together under the leadership of a central person of authority that provided the overview necessary to coordinate a unified effort. These leaders from each crew, be it farming, construction, health care, finances or other functions, would then meet weekly or as needed to discuss and plan, with Stephen in the background, serving as the voice of maturity and experience that helped guide all efforts.

For the most part, Stephen was not involved in the management of day-to-day operations. Designated leaders within the community had the freedom to make independent decisions. That said, it must also be recognized that Stephen would frequently step in and exert his authority to hire and fire, installing or removing someone in a position of power.

Michael O was in charge of the Farming Crew and a widely respected leader within the community from the beginning:

> With the exception of a couple of years, I was on every governing body on The Farm until the time we left in April of '82. The very first one was formed in the winter of 72–73 (aka "wheatberry winter"). It was about 12 people, all handpicked by Stephen. No one from Stephen's family was on it. I remembered feeling really excited about having been on the list of those selected to govern the community, carefully studying who else was on it and then going up to the first meeting to find it was crashed by several people, one couple in particular I remember, who came saying that Stephen must have "forgotten" to mention them at the service when he announced the committee. No one, myself included, had the balls to say anything about it, and the meeting, and governing council, went forward with them on it.

Brandon lived for a short time in those early years with Stephen and his family, giving him the opportunity to observe the behind-the-scenes management of The Farm:

> I can tell you how it ran during 1974. I witnessed and heard family and the usual Farm "heavies" come for visitations to discuss the governance of The Farm. As I recall, Stephen likened it to flying a kite. Everything was discussed, as well as everyone and what they were doing or wanted to do...all ending with what they "should" do. The decisions of who would do what, when and how were then managed into Farm life. Impressions of how someone was faring would be discussed with recommendations for remedial treatments. In today's terms,

one might refer to this as "micromanaging." Stephen would use his family and a few others for some initial feedback on his point of view, and then, his word was God's.

Abbot of the Monastery

While the creation of the community was regarded as a new experiment in social and economic structure, at the same time, universally recognized definitions were often utilized to explain or clarify the community's structure and organization. For centuries humans have gathered together for the purpose of seeking a more spiritual life, familiar in Christian theology as the monastery. The Farm regarded itself as a family monastery with Stephen at its head or as the abbot in charge. Each person joining the community made a personal agreement with Stephen to accept him as their spiritual teacher. Although

For the first 10 years of the community, every Sunday after an hour of meditation, Stephen Gaskin would speak at a gathering of Farm members.

the teachings were unwritten, there was a basic understanding about the type of behavior that was acceptable and general agreement on key concepts such as nonviolence and essential oneness of humanity with the universe.

As Stephen's role as the community's "guru" (a term he never used) became firmly established, the lines between teacher and cult leader began to blur. In the minds of many, he became regarded as an enlightened being, channeling life-force energy referred to as "spirit." Each person had their own interpretation and way of rationalizing their relationship of teacher and student, which to greater or lesser degrees became a form of celebrity or even idol worship.

Marian came to The Farm in the early 70s:

> When I asked Stephen to be a "soaker" (a person granted an extended visit to help them make up their mind about joining The Farm), he said that he was the ultimate authority, that what he said goes. And, around the same time, while he was holding forth to a room full of seated Farm members, looking out at all those worshipful faces looking up at him did give me pause. "Stephen says" was heard frequently for the first number of years. I would say that it felt like a guru-led monastery to me, especially as there was no democracy in the earlier days.

To help explain the relationship between teacher and student, Stephen pointed out that in our modern culture we often have unacknowledged teachers and mentors influencing our direction and attitude in life. From this perspective, he proposed that you could make better choices and put your energy into someone who represented positive and moral ideals without compromising your personal integrity and free will.

Minister or Father Figure?

Again, one important factor that helped establish Stephen in his role was difference in age. When The Caravan landed in Tennessee, most members of the community were in their early 20s, with a few approaching, or over, 30 and only a very small number 40 or older. In contrast, Stephen was in his late 30s, with a much greater life experience behind him. This simple fact of maturity gave him the ability to help steer the direction and make decisions from a broader perspective. Although the term "father figure" may be applied with some reluctance, nevertheless the influence of his more advanced years was undoubtedly a key factor in the acceptance of his position and decrees.

Stephen's primary interface with the community was through the Sunday Service, which in many ways followed a structure familiar to the audience of youth that had grown up attending weekly church services of one faith or another. After an hour of meditation, Stephen would perform marriages and then "gather the flock" in a circle around him. His talk each Sunday could be compared to a sermon and his role to that of a minister or rabbi. Throughout the rest of the week, those seeking guidance could go find Stephen for one-on-one counseling, or come as a couple if advice was needed to resolve marital difficulties.

As The Farm's population grew and its operation became more complex, Stephen had less and less to do with its day-to-day operations. The many different work crews determined their own priorities and implementation of tasks, as outlined by the pyramid of crew chiefs in the various incarnations of governing councils. Stephen was regarded as the spiritual guide pointing the way, while the population of The Farm was expected to figure out how to manage the community's growth and development.

Throughout its history, The Farm has always had a central governing body or board of directors to manage its affairs. However, for the first decade, directors were chosen either by Stephen or from other members of the internal government, not by the community through any type of democratic process. In general, no one really questioned this or saw it as a problem because, for the most part, people were placed in these positions because of their knowledge, skill or demonstrated ability to facilitate people working together.

Social Position

While in the early years Stephen Gaskin did have the power to make and implement decisions for the community, for its daily operation and overall planning, The Farm also operated like a consensus model. The spirit of cooperation meant it was recognized that every person's viewpoint was valid and could contribute to the group conversation.

During this period, Farm residents lived in communal households of up to 40 people, consisting of several families with children, along with quite a few single people. House meetings to organize the tasks at hand brought them together to make joint decisions, whether it was to create a schedule for washing dishes, childcare or getting firewood or who to accept as new members of the household.

It almost goes without saying that The Farm regarded equality as a human right. Stephen defined the roots of racism as thinking your kids are better or more important than someone else's. To ensure that all members held the same rights and privileges with The Farm's internal society, constant peer review would let a person know if they were assuming "social position" or an inflated view of their personal status or importance.

Question Authority—Unquestioned Authority

However, as the person clearly in charge of the monastery, Stephen's power to affect the direction and decisions by the community was unquestioned. On his say-so, any decision by people lower down in The Farm's hierarchy could be reversed. Projects could also be initiated and the resources needed to implement those projects commandeered simply by his expressed desire to see something happen.

For example, in 1974 The Farm's relief and development organization, Plenty International, was founded after Stephen introduced the concept at a Sunday Service. Within a few months, the non-profit was established to serve as a channel to express the community's desire to make a positive difference in the world. Over the next several years, considerable resources, both financial and human, were dedicated to fulfilling this vision.

Plenty's work came to define the best of what The Farm was trying to accomplish. It symbolized the community's true purpose, and everyone in it felt connected to the Plenty projects taking place in the US and abroad. At the same time, while directing attention and considerable resources toward this effort, the community simultaneously neglected its own needs and the development of its infrastructure, resulting in year after year of substandard living conditions for residents.

Expanding on the idea of international outreach and no doubt inspired by the early campaigns of Greenpeace, on another Sunday, Stephen proposed that the community purchase a freighter to transport people and relief supplies on the open sea. Inspired by Stephen's pronouncement, a group of volunteers and their families moved from Tennessee to Mobile, Alabama, to earn money to buy a suitable vessel. After more than a year doing research and attempting to amass funds by working in Mobile's shipyards, they returned to Tennessee and the project was abandoned.

Starting from the early days in San Francisco, Stephen's talks and lectures were recorded, transcribed and edited down into books, records and tapes that were distributed in a number of ways. The Farm's first real business, The Book Publishing Company, was originally established as a mechanism for disseminating Stephen's philosophy and social observations nationally and internationally.

By the late '70s, this publishing business had achieved some success by expanding their list of titles to include books inspired by various aspects of life in the community, most notably a collection of birth stories, *Spiritual Midwifery*, and a collection of vegetarian recipes, *The Farm Vegetarian Cookbook*. However, in the late '70s, when Stephen announced his plan to release a hardcover of his edited lectures, the team of 60 or so people working at The Book Publishing Company rallied behind it with all their attention, pulling energy and resources away from more lucrative efforts.

Unfortunately, by that time, the counterculture movement as a whole had begun to fade, and there was less interest in spiritual philosophy, as the youthful energy of the '60s was being absorbed back into mainstream culture. After years of relative obscurity living in the hills of Tennessee, Stephen and his style of homespun spiritual teachings were all but forgotten. After considerable investment of both time and financial resources, the book did not sell and was an economic drain on the company and ultimately the community.

Working the System

To keep The Farm moving forward, people in charge of some aspects of its operations learned how to work the system to further their goals. With money always in short supply, it became important to win favor with those controlling the purse strings in order to gain access to funds.

As manager of the farming operation, Michael O became skillful at knowing just where or who to nudge in order to gain support:

> Many of us played the "system" and consciously and sometimes craftily manipulated it, even as someone in a position of power. I became the consummate schmoozer of the bank ladies, and courted members of Stephen's family for their interest in nutrition and gardening.
>
> In the 1976 farming season, there was strong opposition to the huge expansion of the farming operation—led by some members within the Farming Crew and other people on the Board at that time. Stephen breezed into one meeting and told those guys that Michael was "carrying a piano on a tightrope" and to back off of him. So I took my "ends" of a worker-based egalitarian, agrarian community over the "means" of having the decisions being made by some sort of democratic process or, at least, free and open discussion. It turned into a colossal error.

Inexperience, combined with a freeze that year that devastated a large cash crop (vegetables being grown as a business venture), resulted in huge financial losses for the community. It was but one more example of how Stephen's authority could be used to override concerns or push forward agendas, sometimes with disastrous consequences.

The Elders

Toward the end of the 1970s, an effort was made to formally recognize those within the community that embodied integrity and credibility to serve as "elders," people who could be sought out for spiritual guidance, help resolve disputes and engage in determining the direction of the community. From the 1,000 or so living on The Farm at that time, representatives were chosen through the community's first democratic vote. There were no candidates or campaigns. Residents were asked to simply list the ten people that best exemplified the role of elder. Everyone was encouraged to not use age as a specific criteria to define the concept of elder, so that the resulting group would represent the broader cross section of the community, including teenagers.

The 40 people who received the most votes were announced at the next Sunday Service. It was no surprise when Mary Louise, one of the midwives respected by everyone in the community for her fairness and compassion, received the most votes. All in all, people seemed satisfied with the results, a true reflection of the people within the community who carried the most respect.

Ultimately the effort did little to alter life on The Farm. The group had no real authority or mission. They were not expected to serve as a government, but were regarded more as spiritual counselors to influence the general direction of the community. Almost as quickly as it was conceived, the elders faded back into the fabric of The Farm.

Carol N was one of those chosen. "I was on the Council of Elders for a while. I know we did the best we could at the time. I think it was an attempt at giving others some responsibility for decision making."

Another elected elder, Albert B, remembers:

> There was very poor control of meetings and agendas in those days. We were accustomed to a charismatic leadership model,

and had little awareness of tools such as facilitation, consensus or conflict management. Meetings went on for hours and hours, gave everyone a headache, resolved very little and were inevitably doomed to be repeated again a week later. There was no filter for what was an appropriate item to be decided by a 40- or 70-member group, and so we would spend weeks deciding on a particular water heating system for a public building or a marital argument that may have occurred in the Florida Farm. Instead of Stephen micromanaging these issues, a group of 70 people tried to micromanage them.

Feeling the Strain

Stephen's ability to commandeer the community's financial resources became evident once again in the early 1980s when he announced that he and an entourage would be making a tour of Europe and Australia. Over the course of the '70s, The Farm's primary recruitment tool for attracting new members was to send Stephen and the community's official rock and roll band (The Farm Band) on the road for a series of free concerts, which included lectures by Stephen. After each national tour, as many as a hundred or more new people would arrive, bringing in fresh energy, donating their vehicles and limited (or even not so limited) financial resources to the community. In accordance with the collective agreement, a few large inheritances and trust funds were swallowed up to provide the money needed to meet the monthly expenses of supporting the growing community.

With the population of The Farm at now more than 1,000, those attempting to manage its finances were beginning to feel the strain of providing for the community's basic necessities. The overseas trips cost thousands of dollars, money that the community did not have. There would be little direct benefit in the form of new members. Many people, especially those managing the community's finances, began to question the wisdom or reasons for such tours. But with no real structure in place for a community decision-making process, their concerns were simply shut down or dismissed.

Of course, after more than a decade as the established authority figure, it was not entirely necessary for Stephen to make his arguments in person. Most residents were willing to follow his direction, with those closest to him able to wield influence on his behalf.

Warren explains:

> Each one of us had a part in that. If enough of us had stood up and said things had to change, we could have demanded more democracy. But we didn't. A lot of fine folks came through there and saw the problems and pointed them out, and when they saw that change was not an option, left to go live somewhere else.

Susanne, a single mother living in the community at that time, remembers: "Many of us found ourselves dumbfounded to see that certain persons with more clout and authoritarian personalities would push others to agree. That was horrible."

Michael O adds, "It is not really fair to blame the lack of democracy on the Farm on Stephen. It took the community's agreement and a strong element of peer pressure."

The Task Force

By 1981 it was starting to become obvious to those managing the community's finances that the amount of income being generated was insufficient and unsustainable. For most of the last decade, the principal source of money coming into the community was being earned by men doing construction work in the nearby towns. However, the first oil crisis, along with double-digit interest rates at the end of the 1970s, had put the construction industry in a tailspin. The Farm could no longer count on the 60 to 70 "basic budget boogie boys" to support its population of over 1,000. Although a number of different small businesses had been launched, none were making enough money to pay real salaries into the community budget in relation to the workers at each enterprise.

To address the situation, The Farm created a task force of community leaders and its top business people to develop strategies for generating more income. Their solution? Tree planting.

The Task Force initiated a series of brigades, sending teams of people throughout the North and South to plant trees. An energetic tree planter could earn as much as $100 a day, top pay in those economic times. A few folks were physically up for the grueling work and became top wage earners, but most were not suited for such intense, demanding labor. After about a year, with little to show for the community in the way of real income, the effort was abandoned.

The Tipping Point

For many people, the tipping point in The Farm's destiny and the breakdown of Stephen's authority took place on a Sunday morning in 1982. During winter months, when the weather was unsuitable to gather outdoors for Sunday Service, Stephen would address the community through an in-house cable TV system that connected about 60 percent of the households. He would talk for about an hour, delivering that morning's "sermon" and then take questions from people calling in via the community's private phone system.

Recently returned from a visit to Plenty's current outreach projects, Stephen gave what was generally regarded as a pep talk, with the goal of inspiring the community to continue its outreach and public service.

Rupert worked as a mechanic and was well liked for his great sense of humor. From his observation:

> When the wheels began to come off, with the community enduring debt, over-crowding, and a ridiculously low standard of living, occasionally one of our braver members would stand up at Services to suggest that we take better care of ourselves (aka democracy)—usually this was in response to rumors of a new

> inheritance or chunk of money coming our way. But Stephen would always guilt-trip us by saying we had to continue being selfless, pouring more and more energies into third world projects and "taking on more" (needy souls) on the home front—in other words, we shouldn't spend the new money on Tennessee Farm projects that might ease our sanitation problems or otherwise make daily life less a grind (particularly for women). We were told we were a beacon, an example the rest of the world looked to, and we had only gotten to this exalted point by being selfless (working like donkeys), so if we started spending more on ourselves we wouldn't be The Farm anymore. At this point, Stephen would kind of look around, but usually no one would rebut his take on things. And I have to say that I bought his argument every time.

But on this particular Sunday, someone did take a stand. A call came in from Michael O expressing concern over the state of the community. In his view, the community needed to do as it had in previous years and close its doors to new people for an undetermined amount of time until the community's finances could be stabilized and improvements made to its housing and infrastructure.

It was perhaps the first and only time Stephen's authority had been publicly questioned. The rebuke carried even more weight because, as the person in charge of the community's farming operation, Michael ("Chairman Mo") was himself widely respected as a charismatic leader. Stephen's infuriation became evident the next morning. He arrived at the morning meeting of the Farming Crew wearing a pair of cowboy boots, a not so subtle symbol of a showdown. The ensuing discussion was pretty much one-sided, with Stephen coming down hard on Michael for insubordination.

Word of the incident spread quickly throughout The Farm. Stephen had crossed the line, expressing anger and using intimidation, both considered unacceptable behavior. It clearly illustrated that Stephen was not infallible; he was simply a man who had become accustomed to power and unquestioned authority. The emperor, or in this case the abbot, had no clothes.

Within a very short time, Michael O and his family left the community. Dozens and then hundreds followed in his wake. Over the next two years, The Farm's population fell from over 1,200 to about 700.

For those leaving, the dream was over. The Farm was seen as a sinking ship about to crash on the rocks.

The Board of Directors

During all this time, the legal framework of The Farm was based on a status established by the IRS for institutions, such as Christian monasteries, called a 501(d). Like a corporation, as a 501(d), "The Foundation" was required to have established officers in the roles of president, secretary and treasurer, and those serving in these positions had been members of the Task Force charged with solving the community's financial crisis.

As part of dealing with the crisis, the elders had formed a committee to reexamine the rules and bylaws of The Foundation. By 1983 the informally organized Constitutional Committee had redrawn the structure into a more formal system with a managing Board of Directors comprising individuals fully empowered to take whatever steps were necessary to save the community from financial collapse.

One of the Board's first steps was to do a full accounting of the community's indebtedness and financial obligations. Under the relatively unstructured communal system, the different working entities within the community managed their own finances, which included the freedom to set up bank accounts and even take out loans from local banks. The Farming Crew had borrowed money to purchase equipment and finance their failed business venture. Each different, new small business had its own accounts and had also borrowed money for start-up costs.

The Farm Clinic had accumulated thousands of dollars in debt with a number of hospitals. Of course, The Foundation had its own checkbook, the account used to purchase food, clothing and on and on, to pay all the running expenses of the community. Accounts with suppliers for food and other necessities had fallen behind and represented another source of debt.

The Board designated a man on the Farm Legal Crew as Farm auditor. He methodically examined the books of every business and made a number of discoveries. The Farm Clinic had accumulated nearly $100,000 of debt with hospitals. There were sordid tales of failed restaurant ventures in Nashville and California,. The Book Publishing Company was selling so many of its titles at prices below actual cost, even without calculating labor, that it lost money on even its most popular books.

The truth was both shocking and frightening. The total accumulated debt was as much as $600,000. Bank interest charges were running in excess of 20 percent. It has been estimated that those debts compounded until they reached $1.2 to 1.4 million by the mid-1980s. With the community unable to even meet its basic weekly operating costs, the Board of Directors concluded that the communally organized economy did not put sufficient responsibility on the members.

In September 1983, an All Farm Meeting was called, and the Board made an announcement. From then on, The Farm Community would no longer pay for any living expenses. In addition, each member would be required to pay in weekly to cover the operating costs and put money toward the repayment of the many different debts accumulated over the previous 12 years as well.

The Birth of Democracy

At the same time, this single act, known by members of The Farm as "The Changeover," transformed the community into a functioning democracy. Every aspect of the community would be analyzed, evaluated and voted on by the members to determine if the expense was essential for its operation or a luxury it could no longer afford. Many services that for years had been supported financially by the community now had to pay their own operating costs, including salaries.

Notice to the IRS

One of the first steps that the community had to take was to notify the IRS that The Foundation was no longer functioning as an income-sharing entity, meeting the IRS's qualification for status as a 501(d). A trip to Washington, DC by The Farm's lawyers, and quick backpedalling by the financial team along with a refiling of the tax reports, kept The Farm from amassing an even larger debt, one that carried the prospect of tax foreclosure. In the meantime, the lawyers restructured The Farm as a non-profit membership corporation, but one without a Federal tax exemption, using the same name as the previous 501(d), The Foundation.

This change helped define the new relationship of people who were members of The Farm and its management, or in essence, its government. Although the largest asset, the land, was set up in a separate trust, the responsibility for the management of all other assets fell to the new corporation's Board of Directors. Permanent residents of The Farm became members of The Foundation and were (are) regarded as shareholders, co-owners of the community's assets. As a member-based organization, the new corporation was required by law to establish bylaws outlining the rights and responsibilities of members, including selecting the Board of Directors. The articles of incorporation defined the length of the terms each Director would serve and detailed the process of selecting new people to the Board, a democratic vote by the members. Several on the Board who initiated The Changeover had plans to leave the community. An election of a new Board of Directors selected through the democratic process was set for the beginning of 1984.

Membership

The Changeover also made it necessary to redefine how a person becomes a member of the community. The loosely defined relationship with Stephen as a spiritual teacher was no longer a relevant factor or criteria for membership. New members joining the community would become stakeholders in the new corporation. However, in the first years after The Changeover, there was no immediate pressure to work out all of the details on this. With the community still in a state of turmoil following the radical restructuring, it was not accepting new members. People continued to leave by the hundreds. By the mid-80s, the population had fallen from around 700 to 250, or approximately 100 adults and 150 children.

All adults (anyone over 18) who had been members of The Farm from before The Changeover were grandfathered in as members of The Foundation. They now had the responsibility of deciding who would be accepted as new members. A consitutional committee was created to establish bylaws that would outline the rights and responsibilities of the members. These included standards of behavior and the process for confronting someone who had violated the bylaws to the degree that their membership status could be revoked.

When The Farm was founded, Stephen had established the concept that the community would be based not on rules but instead on shared agreements. The idea behind this was that rules were set by an established authority, which conflicted with the general attitude of hippie philosophy that rules were meant to be broken. Agreements meant that people were choosing to act from their own free will, voluntarily accepting and cooperating with established community standards. However, as the years progressed, many of these agreements became rules that ultimately reduced the amount of personal freedom and ability of people to make their own decisions.

The core principle that everyone agreed to was nonviolence. As a spiritual community, The Farm had decided that violence of any kind was unacceptable, including anger and intimidation. The new bylaws stated clearly that no weapons were allowed in the community. Anyone who repeatedly used anger and intimidation could be called before a disciplinary inquiry by the Membership Committee. If such behavior continued and a person refused to seek counseling or make any effort to change, membership could be revoked by a vote of two-thirds of the community.

This same two-thirds vote also became the bar for accepting new members into The Foundation. However, if someone does not achieve the number of votes needed to achieve status as a full member, they are not required to leave. The vote simply illustrates that some members do not feel they know the person well enough to vote in their favor. A person may reside in the community as a provisional member for several years before making the transition to full membership.

The Budget

In many ways, the biggest effect of democracy coming to The Farm was empowering each individual to have a voice in the community's operating budget and expenditures. Every fall, meetings were held, and items to be considered for the budget were voted on line by line, with each proposal requiring a simple majority based on a quorum of voting members in order to be added into the following year's budget. The total dollar amount of the items that received a majority vote was then divided by the number of the members, determining the membership "dues" each would be required to pay annually or each month.

The Farm holds quarterly meetings that give all members the opportunity to express an opinion and have their voice heard.

Managers of various community functions and services were required to develop a budget that would be put before the community for approval. For example, it was obvious that the community needed to keep its water system in operation. It was understood that the water manager was seeking only a reasonable amount of compensation for their time and that the other costs for the electricity to operate the pump, the purchase of chemicals and other incidentals were simply expenses that had to be covered. Still, the community went through the process of voting this position and its corresponding expenses into the budget to affirm and acknowledge the shared agreement to pay for this service.

To Pay or Not to Pay: What Are the Consequences?

With the new economic system, each adult member was now required to pay a fixed amount per month into the community, starting after The Changeover at $135. Under the old system, those people employed outside The Farm doing construction or some other work had their paychecks turned over to the community "bank" or finance managers. Now these people kept their paychecks and were responsible to pay monthly dues directly to The Foundation. Companies inside The Farm now had to start paying their employees. Many people were forced to look for jobs outside the community. Some were able to adapt to the change rather quickly, while others struggled to get on their feet financially. If a month went by and an individual or family was unable to earn enough money to pay dues, or for personal reasons decided not to make a payment to The Foundation, this would show up in the community's account books as a debt.

As the months (and years) went by, a number of people began to accumulate an increasingly larger amount owed to The Foundation, and this began to raise questions. In a sense, the community dues were collected through an honor system. What were the consequences to someone that fell behind in their dues payments? What enforcement mechanism could the community use that would penalize people who did not meet their financial obligations?

Shortly after The Changeover, one family announced plans to leave and refused to continue paying the monthly dues. However, since they were continuing to live in the community and use the services provided by The Farm, the Board of Directors felt this was unfair to everyone that was honoring their commitment to cover the community's expenses and payments toward the massive debt. In an unprecedented move, the Board used the community's lawyer to put a lien against the couple and froze their bank account. The decision obviously had some amount of support from a number of community members or it would never have taken place. From a community public relations point of view, this move was a colossal error.

While perhaps freezing the couple's bank account could be rationalized, the move came across as heavy-handed and was very unpopular. Ultimately the Board withdrew the lien, and after some months, the family left The Farm.

You Pay, You Vote

One of the primary privileges for members of The Foundation under the new democratic system was their ability to vote on the annual budget. With a relatively small number of members (around 100), each vote did carry weight and some amount of power when deciding what would be funded for the coming year. However, since not all people were contributing, those who were paying began to feel it was unfair for people not meeting their financial obligations to, at the same time, be involved in the decisions regarding the community's finances. After all, these people could vote to fund a project and then not pay for it. The Board decided that anyone who fell behind in their dues payments by three months or more could not participate in the budget vote.

The decision had the desired effect. Often right before a community vote, those delinquent in their dues would come into The Foundation office and pay the money they owed. Another way they could get special permission to vote would be to agree to a payment schedule with The Foundation in order to catch up and stay current.

While the voting restriction did provide consequences and an incentive to meet the dues obligations, people who were unable to pay began to feel the new economic system was becoming taxation without representation. People with larger incomes could make the decisions regarding what people of lower incomes would have to pay for in the coming year. The decision about whether someone who was behind in their dues could be excluded from voting was finally brought to a community vote. The majority was in favor of maintaining the pay-to-vote status, no real surprise since those voting were also the people who were current with their dues. Over the next two years, the issue was brought before the community three separate times. Finally, by the third vote, in the interest of cooperation and to ensure everyone felt included in The Farm's decisions regardless of their income level, the policy was rescinded.

At the same time, it was also decided that a person could not lose their membership or have it revoked only for financial reasons or debt. The community's bylaws outlined the conduct and behavior that was expected of members, and it was felt that these were the primary matters of importance regarding membership and that it would go against Farm philosophy to ask someone to leave or take away their membership because of financial hardship.

Faults in Majority Rule

The Farm operated in this manner throughout the '80s and '90s, with budget meetings held every fall followed by a community vote to determine the budget for the year ahead and, by default, the amount of monthly dues for each person. Some years, certain budget items were fixed payments carried over from the year before. For example, a massive road improvement project with a budget of $40,000 was financed (using money from the community's savings account) and paid back over four years. After being approved in the budget, it was not necessary to vote again on this for the following three years. A

$70,000 budget for a new water tower was financed the same way for a period of seven years. These parts of the annual budget became fixed and non-optional until the loans were paid off.

It also began to feel redundant to vote every year on other budget items that were no longer seen as optional. For example, a bookkeeper or accountant was an essential service for the operation of the community. The water system had to be maintained and in compliance with state standards. The community maintained a liability insurance policy. These and other items with fixed costs were eventually recognized as non-optional budget components. The Board submitted the idea that only budget proposals for special projects or non-essential services within the community, which may vary from one year to the next, would require a vote.

Coming into the '90s, most Farm members had achieved some degree of financial stability, although members had different levels of income. Some of the businesses operating on The Farm may have become well-established, but were only able to pay modest salaries. A person who had developed professional skills might be doing fairly well, while a young person just starting out on their own might still earn little more than minimum wage. Young families face the many expenses that come along with raising children and won't have the same level of discretionary income as an older couple whose children have left home. Every year, a few individuals or a family had to endure an unexpected financial burden, such as extra medical expenses or the inability to work due to injury or illness.

When it came time for a budget vote, a quorum of voting members were required to cast a ballot for the vote to be legitimate. For 100 eligible voters, a quorum would be 75. Budget votes were passed by simple majority, which meant a proposal might need only 38 voters in order to pass and there were generally always enough people supporting any proposal to produce a majority vote. This meant that the total monthly dues were always the total of all proposals, no problem for people with good incomes, but a strain for those with more expenses or low incomes. Some were also beginning to resent having to pay for services that they never used or projects that did not personally benefit them. If emotions ran high or an issue was particularly sensitive, a few disgruntled members might protest by "going on strike" and refusing to pay.

The result was that, for a variety of reasons, often people would fall behind, accumulating debt to The Foundation, and The Farm was on its way to creating a sub-class of debtors.

It became necessary for the Board to modify the annual budget to compensate for these losses, which over time amounted to tens of thousands of dollars. Although there might be over 100 people eligible and expected to send dues into The Foundation each month, the budget was based on a total of 85 people paying, with the understanding that, from one year to the next, approximately 15 percent of the total population would fall short for one reason or another.

The Pledge System

As The Farm transitioned into the next century, the Board of Directors and a group of volunteers serving on a Finance Committee began to consider a new budget model that could alleviate the economic pressures for those on low incomes, provide a solution for those with grievances, but still make it possible for special projects and services to get the necessary financial support. Outlined and discussed at a number of community meetings, the new budget model was put to a community vote and passed.

Under the new Pledge System, every person would be required to pay a minimum amount that consisted of all non-optional items. The list was expanded to include budgets for aspects of the community used by everyone, such as the maintenance and improvements of roads and public buildings. Beyond this minimum, the budget items vying for pledge dollars would include all projects and proposals, their total costs and the monthly amount per person if everyone contributed. Individuals were then given the opportunity to write in how much they would pledge in support of each project. It could be zero or go beyond the suggested contribution, which meant that the people truly supporting a specific proposal could pay more to compensate for the people who wrote in zero.

At the bottom of the budget paper, each person filled in their total pledge, which included the non-optional figure, and then signed the form, in a sense, establishing a contract with The Foundation. Unlike the previous budget vote with private ballots and majority rule, under the Pledge System, by signing their name, the individual was expected to take responsibility and honor their commitment to pay the pledged amount.

In all honesty, the Pledge System has had mixed results. If one considers the big picture, it has been successful, with the community able to meet its annual budget demands and even expand on the number and variety of projects. In fact, the total amount of money collected is greater than under the previous line item budget vote.

This transition from democratic vote to personal pledge demonstrates that psychological factors are intertwined with the material results. Because each person is given the freedom to determine how their money is spent, the new system has alleviated the concerns and grievances of those "on strike" or simply made people believe that they can truly feel good about what they are pledging to pay. Those who can only afford the minimum are not forced into debt or resented by those paying a higher amount. Instead of feeling like they can't afford to pay for what others voted for and therefore paying nothing, in general people will always at least pay the minimum. Ultimately, this means the total number of people contributing is higher than ever before.

The Pledge System also eliminates the separation between the established permanent members and new people categorized in status as "residents" or "provisional members." Everyone pays the same fixed amount and has the option and opportunity to pledge beyond the minimum.

Virtually every pledge item receives some amount of money, and when possible, the Board looks for ways to fund the entire amount requested. If the allocation falls short, the person or people behind a project must make adjustments. In some cases, they may find it necessary to keep an item on the pledge ballot for two years or more in order to accumulate the entire amount needed to undertake their project or proposal.

Consensus

Consensus is a group decision-making process, often deemed by its proponents as superior to a democratic vote. A voting system results in winners and losers, with sometimes as much as 49 percent dissatisfied and unhappy. In a community, this amount of disagreement can foster factionalism and create obstructive hurdles when attempting to fulfill the mandate voted in by the majority.

When using the consensus model, a moderated group discussion allows every point of view to be heard. Elements of common ground are identified, and attempts are made to reach a unified agreement that satisfies all concerned. Under the classic consensus model, one person who does not agree has the power to block consensus, and the discussion must continue until their concerns have been dealt with to their satisfaction. This individual also has the option to register their concern but agree to stand down and not block the group from reaching consensus. You might say they reserve the right to say "I told you so" should their concern turn out to manifest itself and present a problem later.

The downside of the consensus process is that it can be very slow. Some call it the "tyranny of the minority," in that a small segment of the population has the power to block the majority from moving forward. The amount of compromise necessary to reach full agreement between opposing arguments results, some say, in decisions that become watered-down versions of the original vision. However, most generally recognize that the strongest decisions are made when everyone is in full agreement, and whenever possible, full consensus is seen as the ideal.

Over the many decades at The Farm, it has been found that the range of opinion is like a belly button (or other distinctive part of the anatomy): everybody has one. When dealing with a population as large as The Farm's, reaching 100 percent agreement on any issue becomes virtually impossible. Reaching an informed decision can require making an effort to research a proposal and become truly knowledgeable of all its aspects and implications. With demanding careers and family situations, many people just do not have the time to become educated or participate in lengthy meetings and drawn-out discussions. This does not necessarily stop them from expressing that uninformed viewpoint during a community meeting and blocking the move toward unified agreement.

A Network of Committees

One of the primary ways The Farm has developed to facilitate decision making on key issues or common aspects of community development is through the formation of committees. Because the committees are staffed with volunteers, the core members are passionate and (hopefully) knowledgeable about the issues that the committee is asked to address. Because these volunteers are not elected by the community, committee decisions are recognized as recommendations to the Board of Directors, who can accept, choose not to accept or bring the matter to a broader community discussion and, if required, a vote.

Ideally committee members are able to reach decisions or recommendations by consensus. Their presentations to The Farm community during regular quarterly meetings outline the important points and the research behind decisions or proposals, informing the greater population. Digital communications like group email lists are also used to engage the community in discussions in order to educate or to gauge popular opinion. The community relies on the expertise or greater knowledge of committee members, and if they are able to present convincing recommendations, these are accepted and implemented.

Still, at times, a committee may consist of members representing two sides of an issue, both equally convinced that they represent the best approach and support of the majority opinion. If the committee is unable to reach consensus, a vote within the committee takes place and the results of the vote presented to the Board and to the community. Depending on the issue, it may be resolved by a community vote or tabled for further discussion until a solution becomes clear.

Because the Board of Directors is an elected committee, its decisions have the power to be implemented without a community vote. It has been purposely structured to consist of seven people, in order to facilitate a tie-breaking vote and prevent a stalemate.

It All Works Out

For the most part, The Farm Community accepts decisions made by a majority vote, and those who lose are willing to let a decision stand. Votes of this type usually revolve around large expenditures or purchases and proposals with budgets over $5,000. Those not in favor generally allow the will of the majority to take precedent, understanding that time will reveal the wisdom or folly of the decision. Quite often, many who voted against something become later converts, won over after a project or expenditure has been implemented. By allowing decisions by the majority to proceed with no impediment, community members agree to trust the greater vision represented through the vote. If things do not go as planned, do not flow smoothly or run over budget, again the nay voters will have the power of "I told you so" to use when the subject is reviewed or further discussion is necessary.

Community Participation

In its transition to a democratic collective, The Farm also gave people the freedom to not participate. Members exhibit all levels of involvement, from Type A overachievers to hermits. As one might expect, the greater community consists of the middle ground, with most people involved in one or more aspect of community life beyond caring for themselves. Votes generally receive nearly full participation, and attempts are made to reach each person so that their vote can be included. While there can be elections with multiple candidates for the

Board, anyone who truly desires can find a way to participate in the decision-making process and serve in positions of The Farm's government.

The challenge for the future is The Farm's transition from its founding first generation to the next wave, people who were not from the original movement of '60s idealists. The Farm's next generation is made up of those who were born there and have continued to make it their home and young people of a similar age who have found The Farm and resonate with its ideals. As The Farm evolves, there is a slow but positive shift as positions of responsibility as well as seats within The Farm's government are filled by people representing a younger demographic. It is difficult to predict when that shift will tip in a new direction, when the Board, the Membership Committee and the many volunteer committees will not consist of baby boomers born in the 1940s, '50s and '60s, but the generation born in the 1970s and forward.

This age ratio in The Farm's government is also affected by The Farm's general population demographic. When it comes to filling positions of government, the younger generation is simply outnumbered by the founding members and cannot provide as many candidates. Members of the first generation are no longer raising small children or starting careers and have more free time that can be dedicated to serving the community.

Many variables will need to change before The Farm's government is taken over by the next generation. The community's population also consists of a growing number of another generation, the grandchildren. Just as the important roles of government during The Changeover created what The Farm is today, the decisions by today's Farm will affect their lives and the legacy of the community that the third generation will inherit.

The Farm's history points to the inherent weaknesses of a charismatic leader based organization, but also demonstrates that a community or organization can rise above this, just as The Farm did over 30 years ago. When questioning how a group of intelligent and aware people could be drawn to accept and empower the authority of a central figure, it is important to recognize that most of society gives as much or more control over their lives to their boss at work as members of The Farm did to Stephen. For an example in the greater culture, think Steve Jobs at Apple, messiah and powerful control freak, all rolled into one. Stephen's strength was to help people bring out the best in themselves in the service of humankind. When it became clear this was no longer enough to ensure survival, the people of the Farm took control of the community and changed its destiny.

The Farm's democratic governmental system answers the need in people to be involved in the decision making that directly affects them. While they may not always be satisfied with every outcome, the fact that their voice was heard and their vote was counted the same as everyone else produces a feeling of equality and empowerment that enables the community to continue and progress. A working government is at the heart of a community's stability and its ability to endure.

CHAPTER 2

Earning a Living

NO MATTER WHERE WE ARE or who we are, every society must have ways for its people to generate income in order to function. How this is accomplished is defined, in many ways, by the culture and values of the group. For The Farm, this has been at the core of its evolution and continues to be one of the most significant challenges to sustain its existence.

Going all the way back to San Francisco, the residents of The Haight were basically on their own while experimenting with the early stages of collectivity. Some people worked "straight" jobs like at the post office or bussing tables. Others took advantage of the liberal welfare policies and food stamps program in California at that time. Communes and collectives served as loose structures to keep the cost of living down to nearly nothing, so that a young person could exist on little more than a bowl of rice and music.

When Monday Night Class was transformed into The Caravan, the dynamics had begun to shift. Buses would need gas every day. Buses would break down and need new parts. People living on the road needed food. All this took real money, and as the mobile community began to crystallize apart from the support system that had flowed so easily in the city, the connection between work and sustenance was seen as a vital relationship on the path to enlightenment.

After entering a new city, groups of people would knock on doors seeking employment of any kind. Young, eager and with only a short-term commitment, the enthusiastic workers were able to experience the fruits and rewards of their labors, gaining the understanding that work could be seen as the expression of love and that money could be transformed from the root of all evil into the material manifestation of energy.

Right Livelihood

Settling on the land in Tennessee was an even greater shift. Almost overnight, the vision went from how do we get by another day to how do we build an existence that can sustain us over a lifetime? What does it take to build a town from scratch? What form does this take when you have rejected the established routes as set in place by the greater society?

In seeking the raw essence of what this meant, Stephen put forth the Buddhist concept of Right Livelihood, defined as work that is seamless with your ideals. A person's daily toil should not compromise their principles, but serve as a tool for the expression of their beliefs and outlook on the world.

In the first years of The Farm, each person had unlimited opportunity to discover how they could contribute. With so much need before them waiting to be fulfilled, it seemed logical to begin with the basics and go from there. To feed the people, we would grow food. The community must have water. The buses would serve as immediate shelter, but it was clear that buildings of all types must be constructed. The sick must be cared for. Children must be helped to learn. These elements of daily life became the foundation for what was truly important and thus defined the work needed to be accomplished each day and the goals for the foreseeable future.

In certain ways, the economic structure of The Farm became its great strength and its weakness. The spiritual truth that resonated in all hearts was that we are all one. Under this spirit of unity, we would care for each other, using the Book of Acts from the Bible as a guiding principle that could be understood by all who joined and by the greater society outside The Farm. No one held personal money. All income was shared and used to support the whole. This shared economic responsibility also left the role of earning money in an open and nebulous state.

On the early Farm, it was believed that for a person to find true fulfillment in their work, they should choose a task that resonated completely with who they were as a person. This would allow them to put forth all their energy without holding back, achieving joy combined with material results. People were allowed and encouraged to find the work that suited them best, often trying out various types of tasks and endeavors until settling in on a field or career path that felt right.

Because no one used or needed money for their life on The Farm, the concept of starting or running a business did not enter most people's minds. They worked at tasks that supported the community, and if this also provided a way to generate revenue for The Farm, so much the better.

The Farm's name comes from what was identified as the clearest form of right livelihood, growing food. As all gardeners have discovered, it is easy to grow more than you can consume, and selling extra produce to Tennessee neighbors became a way for the Farming Crew to generate the money needed to buy seed, to put gas in the tractors and cover the many expenses encountered in the formidable task of feeding several hundred people.

Early on, The Farm was introduced to sweet potatoes, a staple of the South. Tennessee's rural population in the 1970s still raised home gardens, and the Farming Crew discovered there was a demand for sweet potato "slips," sprouts from potatoes grown the previous year. This became one of the first commercial ventures on The Farm, grown specifically for income. Unlike other crops that may take many months until harvest, sweet potato slips achieve their state of readiness rather quickly, over just a few weeks. Sold in the spring just after the last frost, they brought in money to The Farming Crew when it was needed most, the busy planting season.

Gaining Skills

During these early years, money earned from working outside The Farm was almost a side benefit to the actual purpose: training. Being green city kids who had never done much in the way of actual work, members of The Farm had to start learning real skills, and quick. In order to gain knowledge about building, some folks rented a house in Nashville, about 70 miles to the north, and got jobs doing construction. They learned how to swing a hammer and basic framing, how to properly construct a roof and lay shingles so that they won't leak. From building the foundation to finishing the interior, every aspect of construction had to be learned by someone so that those skills could be brought back to the community and used to build our town. The Nashville house became a small commune of its own with a revolving set of families and workers, as people became anxious to acquire the skills through this hands-on apprenticeship.

The Book Publishing Company

As a way to disseminate Stephen's spiritual philosophy and outlook on the universe, his lectures at Monday Night Class were taped, transcribed and then edited. This was turned into a book of the same name, printed and distributed by a small San Francisco publishing company called Book People. To capitalize on the new-age mysticism emanating from San Francisco, Random House picked up the distribution rights to *Monday Night Class* and published a second book of edited transcripts from Stephen's talks on the national lecture/bus tour, entitled simply *The Caravan*.

The prestige and excitement of acceptance by this international publishing powerhouse quickly diminished as it grew clear that Random House was not going to put its advertising might behind the promotion of these esoteric titles. Disillusioned and with the youthful bravado that exemplified The Farm at that time, the community took back the rights to the two books after the contract ran out and made the correct assumption that it could do a better job of promotion than a relatively disinterested giant corporation.

The Book Publishing Company was formed as the first real business on The Farm. Assuming it would need to manage every aspect of production, a few folks were dispatched to Nashville where they found jobs in the printing industry. Following this same hands-on approach that had worked so well for learning construction, the people engaged in this work studied the craft, becoming familiar with the machines and the many steps necessary to convert raw paper into the printed word.

Within a few years, The Farm had acquired an array of printing presses, including the equipment needed to put out beautiful full-color covers and interior illustrations. The Book Publishing Company set up the means to produce the aluminum plates used by the presses. It established a darkroom for a team of photographers. Involvement in every aspect of publishing, including an office staff and sales force, gave The Farm a better understanding of how to start and manage the businesses it would need to move beyond skilled and unskilled labor and into a fully functioning community economy.

Farm Hands

In certain ways, The Farm's move to rural Tennessee came also at a good time for the area's farmers. Rural populations were in decline as their youth continued to migrate into the cities, seeking more skilled and better paying work in factories. The Farm filled a void, assisting local farmers as an enthusiastic labor force. The community also saw this as an important aspect of neighbor relations. Members would work with the farmers, especially around harvest time when extra labor was needed. In addition to building trust and better relationships among the neighbors, these workers also provided a small revenue stream for The Farm.

From week to week, The Farm could survive and buy groceries to supplement the food grown by the Farming Crew with the small amounts of income trickling in from field hand work plus the few construction crews working in Nashville and eventually towns closer to The Farm. However, once each year, a larger sum was needed to make the land payment of several thousand dollars.

In response to this immediate need, The Farm did what hippies everywhere had been doing whenever the rent came due: they went to the offices of Manpower, a company that hired out day labor.

Taking advantage of its fat labor pool, The Farm would dispatch vanloads of eager workers flooding the Manpower office with ready and willing warm bodies. Manpower's regular labor force tended to represent the outer edge of society, the uneducated, the unemployable, often alcoholics and homeless. In contrast, Farm workers were bright-eyed and bushy-(pony)tailed, willing to work hard at any task, motivated by the need to support The Farm. This surge would last for about a month, until enough money was earned, and then life on The Farm for the most part would go back to normal.

The potential from this surge of income was not lost on those people managing The Farm's finances. Why not continue sending people out to work, only do it in our own way, where we make all the profit?

Farm Hands was established early on as a day labor company serving towns much closer to The Farm, enabling workers to put in more or less normal hours and return home at a reasonable time each night. Office space was rented in the nearby town of Columbia and set up with a phone and office managers. Once the word got out, the work was fairly steady, and the business became moderately successful. However, its managers faced the daily challenge of finding people willing to work. People preferred to work on a crew engaged in some task or service inside the community and were not necessarily all that anxious to work in town at some job that had no personal relevance and could even at times be somewhat demeaning. For example, a local aluminum smelter hired laborers to load railroad cars with slag, jagged hunks of residue the size of small boulders. Although the company provided gloves, these were shredded in a very short time. The atmosphere was dismal, abundant with unpleasant fumes and a bleak landscape. There were times when it was better to say no thank you.

New Directions

Although established initially to promote Stephen, the Book Publishing Company began to expand its list of titles to include books based on other innovative aspects of The Farm's lifestyle.

Beginning with the first baby deliveries on The Caravan, Ina May and a few other women had been developing their skills as midwives. With The Farm in the midst of a baby boom, hundreds of mothers were giving birth at home, without anesthetics, and in the process, generating better outcomes and statistics than the medical establishment. The desire for a more natural approach to pregnancy and birth was an outlook The Farm shared with its counterculture colleagues across the nation and around the world. To help foster this movement, each woman who had given birth on The Farm was asked to write down and share her experience. This published collection of stories, *Spiritual Midwifery*, became one of the first truly successful titles for The Book Publishing Company. The last chapter was filled with information and instruction that a lay midwife or a family might use to facilitate a good birthing experience. The book became a manual, as well as a fascinating piece of literature, and helped inspire a radical change in the way birth is managed everywhere today.

The second success story was *The Farm Vegetarian Cookbook*. One of the first agreements established after reaching Tennessee was that the community would use no animal products. This was done both for practical reasons and for compassion with other living creatures in the world. Again, this lifestyle choice resonated with a broader movement that was taking place beyond the borders of The Farm. The term "vegan" was almost totally unknown then, but the spiritual consciousness brought about through the hippie movement became like wind in the sails that moved forward this massive cultural shift in the way people eat. Through its size and collective unity, The Farm was able to distill this aspect of the counterculture and put the information back out to help crystallize the trend toward a healthy lifestyle far and wide.

The Big Dummy

The third chapter in the Book Publishing Company's growth and success took everyone by surprise. For several years, The Farm had been using amateur (ham) radios as a communications tool for staying in touch with affiliate Farms around the US and Canada, people in other countries working with Plenty International relief projects and with Stephen when he and an entourage were on the road. The Radio Crew also began using less expensive citizens band (CB) radios (which did not require a license) on The Farm to facilitate communication by the Midwives and the medical staff that served as their backup.

The Farm had also started a trucking company, and its drivers discovered that truckers everywhere were using CB radios to alert each other to the whereabouts of police radar. In the 1970s, in an effort to improve fuel economy and make the country's highways safer, the Federal government reduced speed limits on the interstates from 70 to 55 miles per hour, a move that was quite unpopular among the nation's drivers, especially truckers. These asphalt cowboys were using CB radios to broadcast the whereabouts of "smokies" (as in Smokey the Bear) to warn their counterparts a few miles in either direction or to send out an all-clear signal that all were free to "put the pedal to the metal."

The phenomena inspired a hit song, which in turn spun off a successful Hollywood movie and a national craze. Almost overnight, millions of people were buying radios, installing them in their cars, even in their homes. It seemed like every person had to have a handle (nickname) to go along with their new CB personality.

Two guys on the Radio Crew put together an instructional book that covered all aspects of using a CB radio, including technical information on how to build and install antennas, knowledge that came from The Farm's experience with amateur or ham radios. Suddenly

The Book Publishing Company had a title that went beyond the counterculture fringe and resonated with the mainstream. Its staff of artists filled the book with both amusing and informative illustrations, including a colorful character called "The Big Dummy."

The Big Dummy's Guide to CB Radio hit The Farm like a tidal wave. Instead of sales numbering in the dozens or even hundreds, this book was getting orders by the thousands. Along with this rush of success came some hard lessons about the publishing business.

Wal-Mart alone placed an order for several thousand books. Since The Farm handled all aspects of publishing including the printing, the sale was very profitable, and it felt good to be behind the wheel of a million-seller. Unfortunately, reality struck a short time later when Wal-Mart returned all of the unsold copies and demanded the balance of their money back.

Book publishing is a funny business. In order to keep their inventory fresh, retailers have developed a system where they place an order, take in the inventory, hold it for 30 to 90 days and then return any unsold or remaining inventory for a refund. They then reorder the same book and repeat the process. This minimizes the retailer's risk and investment while generating tremendous waste, with each returned book becoming a second or used book, often trashed.

It was a tremendous shock and a blow to the Book Publishing Company, as well as The Farm as a whole, and a hard lesson in modern capitalism. Every returned book had a Wal-Mart price sticker in the corner. Scrambling to salvage what they could from the mess, the Book Publishing Company put people to work removing the stickers, carefully using a hot clothes iron to soften the glue so each could be peeled away. The books were then sold at a huge discount to a distributer in Australia where the CB radio craze was just beginning.

Food Companies

Because The Farm's dedication to a vegetarian diet was such a large part of the community's lifestyle, it launched a number of enterprises related to this ethic.

The Farm had gained notoriety because of its concentration of people all consuming a similar yet unorthodox diet based on soy. Soy was widely recognized for its high protein and viewed as an integral solution to world hunger. Early in its history, the community was contacted by a university professor researching foods around the world who introduced The Farm to tempeh, a fermented soybean food from Indonesia. In Indonesia, soaked beans were wrapped in banana leaves, exposing them to a wild mold spore. In 24 to 48 hours, the spore would multiply and turn the beans into a solid mass that was then cooked and eaten.

Tempeh spores had been isolated in a laboratory, and some of the first strains were sent to The Farm in the early '70s. Before long, tempeh had become another staple of The Farm diet, yet one more way to for the community to consume soybeans on a regular basis.

While The Farm was one of the first to discover and enjoy tempeh, others in the US were learning about this "new" food, and there was a small but growing demand for pure high-quality tempeh spores. The Farm was encouraged to develop a business around the production of tempeh spores, which required a sterile environment and careful but manageable sterile technique. Within a few years, tempeh shops were springing up here and there around the country, small enterprises producing tempeh to sell in natural food stores. These shops all became customers for a small business on The Farm that produced spores for tempeh companies around the world.

The Farm Market

By the late 1970s, with close to 1,000 people to support, The Farm began developing a number of small businesses in the nearby town of Columbia, which had a modest population of around 40,000 people and a fairly strong economy. The Farm rented a storefront on the main drag with a large warehouse behind and a couple of adjoining smaller storefronts. The Farm Market was like a vegetable stand. Behind the large windows facing the street, tables were filled with produce, some from the fields of The Farm and the rest purchased from wholesalers in Nashville. Business was brisk, and the Farm Market became quite popular, giving local Tennesseans another way to meet the hippies, satisfy their curiosity and take home good food at the same time.

Next door, Skyrocket Electronics served as the counterpart to the *Big Dummy's Guide*, a small electronics shop that installed and repaired CB radios, manned by a few members of the Radio Crew. The Farm's mechanics jumped on the bandwagon and opened High Gear Garage to help bring in some needed funds to the Motor Pool back on The Farm. Everything seemed to be going along pretty well.

About that time, The Farm was given a large flour mill in east Tennessee. The deal was pretty straightforward: Tear down the old mill, salvage and move all of its components and reassemble it as another new enterprise for the community. Since no buildings on The Farm were large enough to house the mill, it would be installed in the back warehouse of The Farm Market, which had plenty of extra space and the height necessary for the equipment. The team worked for months moving and setting up the equipment. Just as the mill was about to begin operation, City inspectors got wind of the plan.

Unbeknownst to the naïve Farm members, milling operations are highly flammable, and can even explode. The flour dust in the air can ignite almost like gun powder from even a small spark. There was no way the City of Columbia would allow the mill to open.

It was like another punch in the stomach, with thousands of dollars and hundreds of man-hours down the drain. It also seemed to coincide with a downturn in the country's economy. The CB radio craze had begun to fade, and the shop in town became less profitable and closed. The auto repair shop had never really got on its feet. Within a few months, the Farm Market had closed as well.

Everybody's Vegetarian Restaurant

When people think of food, they immediately think of restaurants, so it was no real surprise when The Farm decided to open a vegetarian restaurant in Nashville. A building was found in a good location downtown near Vanderbilt University, and the community's construction company engaged in the renovation. In a short while, Everybody's Vegetarian Restaurant opened its doors. Unfortunately, Everybody's also closed after a couple of years, a victim of the wrong place at the wrong time. Perhaps in California the eatery might have stood a chance, but in the early '80s, the vegetarian diet was still considered far left of mainstream, and there simply wasn't a large enough audience in Nashville for such cuisine. Restaurants work on small profit margins, and there is a real reason why franchise chains are able to dominate the food landscape, wringing out profits with their bulk-buying power, along with standardized recipes and labor-saving preparation techniques. Whether due to mismanagement, a constantly shifting inexperienced labor pool or the recession that hit the US economy at that time, Everybody's Vegetarian Restaurant became another business blip on The Farm's historical timeline.

Farm Soy

Early on, The Farm set up a small facility to produce soy milk and tofu for the community, expanding production capability as the population grew. In the late '70s, The Farm started a satellite or branch community in California, opening a soy dairy as the enterprise that would support its members. However, since the group in California had no land and little startup capital, they were forced to rent places to live and a shop for the soy production. Although there was a good market for their products, in the end, sales were not enough to match the cost of living on the West Coast and the "California Farm" was closed, with its members called back to Tennessee.

By the early 1980s, there was enough awareness of soy products in the greater culture to establish a demand for them at the natural food stores in nearby Nashville, allowing the community's Soy Dairy to supplement its tofu production for The Farm with sales to Nashville and a little cash flow.

Farm Foods

Farm Foods was the umbrella company that managed the Good Tasting Nutritional Yeast and tempeh spore businesses. The company's most ambitious endeavor was to commercialize one of The Farm's favorite foods, ice cream made from soy milk, known affectionately (and commercially) as "Ice Bean." There was also a practical reason for the new name. The dairy industry has had a strong political lobby in Washington for many years, and the law books state that it is illegal to use a word like "cream" unless a product contains a specified minimum amount of milk from cows.

In order to ramp up production to an industrial scale, workers with Farm Foods found an ice cream plant in Memphis, Tennessee, that would rent their facilities for limited runs of the soy-based product. A team would head to Memphis about once a month and do their production over a weekend when the plant would ordinarily be closed. It was impractical for The Farm to make soy milk at its dairy facility and transport that to Memphis, so the culinary engineers at Farm Foods developed a recipe based on commercially available soy powder. The texture wasn't as nice as the fresh-made Ice Bean on The Farm, but it had a good flavor, and by the early 1980s, Farm Foods was making inroads for distribution into the country's natural food stores.

The Farm Building Company

The real backbone of The Farm economy throughout its collective period was The Farm Building Company. The men who had learned basic construction skills working in Nashville in the early days were now crew leaders, heading out into the local towns in the early morning hours, each with a vanload of tools and a handful of guys or nail-bangers. Crews began to focus on specific skills. In addition to the basic framing, there were crews that did masonry and interior finish work, as well as electricians, painters and roofers, pretty much a crew for every aspect of construction. At its peak, there were about 75 "basic budget boogie boys" generating the primary income that supported the over 1,000 population of The Farm. Hundreds of barns, homes and commercial buildings have been built by Farm construction crews.

Solar and Wind

During those early years, The Farm would refer to construction work as "Plan B." While building someone a nice home was considered right livelihood, the construction materials and designs did not truly reflect the community's greater vision of sustainability. "Plan A" was considered any work that would be more in line with The Farm's ideals, which included energy-efficient engineering and alternative power generation.

In the late '70s, President Jimmy Carter established tax incentives and funding for a number of alternative energy technologies. The Farm's Solar Energy Works was created to take advantage of this opportunity. This crew began installing solar water heating systems, primarily at homes in the Nashville area. Solar collectors began popping up on roofs all over town.

There was also Federal money for alternative energy projects. The Farm's radio and electronics' crew started a business called Solar Electronics that began acquiring government contracts to design and install energy systems in various parts of the country. The Radio Crew had gained experience setting up towers for communications antennas and was able to use these skills to install 200-foot wind generator towers in the mountains of North Carolina and on remote Federal land in Louisiana. Solar photovoltaic panels were installed to supply all of the power needed for a rest stop along an interstate in Alabama.

When Carter lost the presidential election to Reagan in 1980, all funding was cut, and contracts for these types of projects evaporated. Without the steady income from new installations, The Farm's Solar Energy Works and other solar water installation companies in Nashville could no longer afford to stay in business or maintain the systems they had installed. Over the next decade, the rooftop solar water heaters all disappeared. Without money for maintenance contracts, many years later word came back that some of the wind generators had toppled to the ground.

Alpha, Beta, Gamma

The year 1979 was pivotal for both the country and The Farm. Decades of support by the CIA and other wings of the Federal government for a ruthless dictator, the Shah of Iran, triggered a rebellion in that country that ignited with the storming of the US Embassy. A total of 18 US citizens were captured and held as hostages. Oil facilities in the country were nationalized, setting off an energy crisis that caused gasoline prices to triple almost overnight.

That same year, the Three Mile Island nuclear plant in Pennsylvania went into meltdown, releasing radioactive vapor into the atmosphere. In spite of this early warning, US power companies continued to develop and expand their construction of nuclear power plants across the country.

During this same period, the team at Solar Electronics began using its knowledge of electrical circuits to research the development of products that could be manufactured on The Farm as another revenue stream. The "Fuzzbuster" was a popular new consumer product at the time, a device that could detect police radar tracking speeders on the highways, and like the CB radios, was another response to the reduced speed limits. Back in the Solar Electronics R&D labs, an idea was born. Why not design a "Nukebuster" that would warn drivers if they were behind a truck hauling nuclear fuel or nuclear waste?

Traditional Geiger counters were massive units weighing several pounds, typically operated by the army and civil defense brigades. The units were clunky and devoid of any features. They responded to the presence of radiation with audible clicks that increased in number in relation to the amount of exposure. Surely it would be possible to build something better.

At this time, the Solar Electronics team was also working with an engineer from outside the community to develop a device for doctors and midwives to detect the heartbeat of an unborn fetus. One member of the electronics crew realized that the same averaging circuit used in this fetal heart monitor could be modified to track radiation. This made it possible for the impulses emanating from a Geiger tube to be translated into an electrical signal that could display a response on an analog meter with a needle that visibly illustrated the amount of radiation present. Additional circuitry was added that would set off alarms when the amount of exposure went beyond specified levels.

By the end of the year, the Nukebuster was in production. The crew worked out of two old house trailers parked at the base of the antenna towers next to the pear orchard. About six of them were in one trailer assembling units, testing them for quality control and readying them for shipment. A second trailer housed the office staff and sales force. A third trailer was brought in to give research and development workers their own space.

Making a profit was a bit more difficult. In spite of the Three Mile Island accident and the disaster at the Russian facility in Chernobyl a few years later, American consumers were not concerned enough about exposure to radiation to warrant the purchase of a personal Geiger counter for their car. After all, you can't see, smell or taste radiation. What is there to worry about?

As part of the education process, Solar Electronics learned that the use of nuclear materials goes well beyond generating power at nuclear plants or manufacturing bombs. Most people are aware of the medical uses for nuclear substances, but the applications also include industrial welding and certain types of photography. These users were not going to purchase a product called a Nukebuster.

The name was changed to Radiation Alert, and along with the new name came a more streamlined design. No longer engaged in solar installations, the company also changed its name from Solar Electronics to SE International (SEI). Regardless, sales were still slow. The industry as a whole had not gained the awareness or realized the dangers of exposure to even small amounts of radiation, and there simply wasn't much demand. It would take several years (assisted by additional nuclear and other disasters) before SEI could really get on its feet.

Where Is the Money? Another Chapter of The Changeover

In the year's before and leading up to the Changeover, all businesses on The Farm were expected to earn money and pay salaries, with one catch. Individuals were not paid directly, but each company was supposed to write a weekly check to the community representing the payroll for its employees. However, as is often the case with new startup companies, any money generated over expenses was reinvested in the business. Workers desired new and better tools and equipment. Sales teams wanted funds for advertising and to cover the costs for exhibiting at various trade shows. Virtually every new business was a "bootstrap" operation, with no startup or investment capital, just a lot of sweat equity.

Because these small startup businesses were not carrying their fair share of the load, the overwhelming weight of responsibility to come up with money for the community's operating expenses rested squarely on the shoulders of the guys doing construction work. However, when the first oil crisis and recession hit in the early '80s, this dried up as well. Rising gas prices took an even bigger bite out of the profits, increasing the cost of doing business for all of The Farm's small companies.

Getting Real

Like all other aspects of Farm history, operating a business was entirely different after The Changeover. From that day forward, each Farm enterprise had to pay its employees directly. All at once, the businesses had to meet real payrolls, and this brought reality into focus. With the communal system, although the money paid to The Foundation was supposed to be based on an amount earned per person, the funds would vary from week to week and usually fell far short of even minimum wage. Flush with extra labor, many of the companies had excessively large staffs, often an attempt to make up for the lack of training or skilled workers. After The Changeover, each worker needed real money to purchase food and other necessities. It became clear right away how much money a business was actually earning and which employees were essential to an operation.

Virtually overnight, Farm businesses went through a downsizing process. Dozens of people became "unemployed" and were left on their own to figure out an income. Coming out of the communal system, The Farm's newly unemployed faced additional handicaps. These folks did not own a car or have a dime to their name. Everywhere people were scrambling just to survive, and it's no wonder they began leaving The Farm in droves.

Privatizing

Legally, before The Changeover, all of the many small companies were owned by The Foundation. However, without the knowledge, expertise and motivation of the managers and workers, these companies would not have existed. Rather than force the businesses to close and reopen under new names and under "new" management, it was generally recognized that the businesses would continue operation, but would be privately owned and operated by the principal managers.

If a business had a significant amount of valuable equipment, an arrangement was made with The Foundation to sell that equipment to the new owners. For example, the fellow operating The Farm's heavy equipment purchased from the community a dump truck and other machinery that the business used. Before The Changeover, the satellite dish installation company had taken out a bank loan to finance its startup costs. This note became the responsibility of the new owners.

Larger businesses like Farm Foods, The Book Publishing Company and Solar Electronics became part of a new corporation. Shares of Solar Electronics stock were distributed among its current managers, with The Foundation maintaining the majority. Farm Foods was sold to a corporation from the northeast, and its staff moved along with it. The Book Publishing Company, The Farm's original business, was kept as the only business owned by the community as a whole.

Survival

Lewis County, Tennessee, has long been recognized as one of the poorest counties in the state. At the time of The Changeover, Hohenwald, the county seat, had no major industries, no national franchises and virtually nothing to offer in the way of employment opportunities. Other towns close to The Farm were not much different.

Even driving to find work was a challenge. The Farm was at the end of a long dirt road. It was 50 miles to the nearest interstate highway, and from there, it was another 35 miles to Nashville, the closest urban center. Most of those who went there to find work simply moved off The Farm and up to the big city, unwilling to endure and unable to afford a four-hour daily commute.

Ultimately the people who remained on the land were the ones who figured out a way to earn a living. Those who had already been engaged in a business or job that produced an income before The Changeover were in the best positions. Most members of The Farm Building Company were able to make the transition pretty smoothly. The satellite TV installation business kept going. Other small businesses like the Farm Soy (the soy dairy) and the tempeh spore business slowly got on their feet and eventually were able to support a small staff.

Some people took the skills they had learned on The Farm and looked for employment outside the community. An auto mechanic, a plumber and a few people here and there found jobs, but most who stayed were either part of a Farm business or worked on their own, picking up odd jobs, doing remodeling or small construction projects. Many of the women who had been part of The Farm's Clinic attended a local community college that offered a two-year nursing degree and then found work in nearby hospitals and clinics. A few folks became teachers in county schools. Either you found a way to earn a living, a way to survive in rural Tennessee...or you left.

Earning a Living in Community

After The Changeover and over the next 20 years, The Farm became like a laboratory and an incubator, identifying the different types of work and small businesses that could provide employment in a rural setting. For many of those who remained, the first and biggest decision was really a lifestyle choice.

Typically, people will choose a career path and then find it necessary to look for a job based on that skill or knowledge. The downside is that you must go to where the job is, regardless of whether it is located where you wish to live. The other route is to choose your lifestyle and where you want to live and then identify the types of employment available or develop an economic support system that will function in that environment.

The people who remained on The Farm after The Changeover were making a conscious decision to remain in rural Tennessee, with a daily life surrounded by nature, where their children would be free to roam far away from crime, traffic and pollution. With that choice came the acknowledgement that jobs were scarce and employment opportunities limited at best. Still this core group survived and eventually prospered. Along the way, it became evident that the opportunities for earning a living were much broader than first imagined or were evident immediately after The Changeover. Like anything else, taking advantage of these opportunities required perseverance and motivation. Still, it has been possible to identify what qualities are necessary in an individual, what types of skills are most desirable and some of the common elements that successful businesses on The Farm share.

In many ways, The Farm's ability to survive through difficult times in the country's economy is due to the fact that the income of its members is spread over many different revenue streams. The large population base has meant that its members have developed diverse ways to support themselves, from manufacturing to distribution to a wide variety of services, as well as direct employment for businesses and companies outside The Farm. Smaller intentional communities can suffer if they rely on a single or primary business that supports the group. If that business fails or struggles due to changing demands or a downturn in the economy, the solvency of the community as a whole can be greatly affected.

The Changeover required members of The Farm to determine how to earn a living that best fit their desired lifestyle and the skills they had to offer. The immediacy of The Changeover forced them to identify ways to earn a living that could be initiated quickly and without a lot of investment.

Construction work had always been the primary source of revenue for the community, and the people who had those skills were able to transition fairly easily. Most already owned their basic tools, and it was a relatively small investment to add any that were needed. The main other ingredient they needed to get started was transportation. A van, a truck, even a station wagon can serve as a mobile tool box. The next step was to print up a few business cards and rely on word-of-mouth advertising to keep new jobs coming in. The Farm's own economy has generally been able to employ several people who work almost exclusively inside the community. It has also been interesting to observe that many of highest income earners on The Farm have been those engaged in construction, especially those who have become licensed contractors. These entrepreneurs who have developed the ability to manage people and multiple jobs simultaneously, along with sales and customer demands, have been able to climb to higher rungs on the ladder of success.

Arts and Crafts

Another common means of employment that has worked for some people on The Farm and others living a rural lifestyle is some form of arts and crafts. The country home is an ideal location for an artist's studio, providing the solitude that inspires creative work as well as the extra space it often requires.

Keeping the tradition alive from the hippie times of the '60s after settling in Tennessee, members of The Farm continued to add a splash of color to their wardrobe by making tie-dyed T-shirts. In the early '80s, before The Changeover, a few of the more talented designers started a small company producing tie-dyed shirts and selling them at various craft fairs. The business was able to continue operation after The Changeover, and many people who left The Farm at that time were able to take this skill with them, launching successful tie-dye businesses in many different parts of the country. The skill has been passed on to the next generation who continue to find a market for the colorful patterned clothing.

Mail Order and Catalogs

One of the more common types of businesses that can work in a rural intentional community setting is mail order, which now includes Internet sales. With this type of business, the supplier sits at a central lo-

Operators are standing by to take your order at The Mail-Order Catalog, a small business on The Farm.

cation with their inventory of product. Customers contact them, place the order, and then the product is shipped and delivered. The broad availability of delivery systems like United Parcel Service (UPS), the US Postal Service (USPS) and other specialized delivery services like Federal Express and DHL has made it possible to be situated virtually anywhere and ship product to a national or international customer base. The Farm has several businesses of this type, ranging from the solo entrepreneur to those that employ five to ten people.

One of the best examples is a Farm business that is true to its name, The Mail Order Catalog. It was started by the couple managing the community's Book Publishing Company, Bob and Cynthia. As a publisher, the Book Publishing Company sold its books wholesale to distributors, with the printed pages eventually making their way to the shelves of retail stores. The Book Publishing Company did not sell directly to the end customer, the person buying the book.

The Mail Order Catalog started as a direct mail bookstore selling the titles published by The Book Publishing Company. Being in the business, Bob and Cynthia knew about seconds, the returns that all publishers have to deal with, and they saw this as an opportunity. They began purchasing the seconds from other publishers and offering these books at deep discounts, as much as 50 percent off the list or cover price, to their customers. They had tapped into one of the golden rules of business and sales—people want a deal! The discounted books sold much faster than the titles from The Book Publishing Company, and it drove the business into a volume that allowed them to hire staff. Following this success, they began adding other products that would appeal to their customer base, people interested in a healthier lifestyle. As the business grew, they bought out other companies selling compatible products, increasing their product line to include a wide variety of food items that could be easily stored and shipped without refrigeration. These now make up the bulk of their business, although they still offer the book seconds.

Mail Order Catalog has been under development for over 20 years now. Bob and Cynthia started small, working out of their home, printing a small catalog that they would mail out a few times a year to an expanding list of customers. As the business grew, the number of items increased and so did the size of their catalog. One of the primary costs of doing business became and continues to be printing and mailing. While orders placed through an online catalog and Internet sales have now become an increasingly significant portion of their daily orders, many customers still need to receive a catalog in the mail and in their hands, to stimulate and motivate them to place an order.

The Mail Order Catalog has three major components: managers who supervise staff, sales and inventory; people who take calls and enter orders into a computer; and people who take care of shipping. Internet orders that come in during the night are entered into the system each morning. A large part of their business is tracking and maintaining or restocking inventory. By monitoring which items are selling the most, they can spot trends and ensure that adequate supplies are on hand in order to fulfill orders quickly.

Other Mail Order/Catalog Businesses

The tempeh spore business discussed earlier is also essentially a mail order business. It consists of the owner, who originally learned the production techniques, along with a few part-time helpers. The core of this business is maintaining pure strains, with constant vigilance to ensure there is no contamination, rogue bacteria that would grow something other than tempeh when added to the soy or various grains on which commercial tempeh is grown.

In a way, Mushroompeople is similar to The Tempeh Company. Mushroompeople is another small mail order business located on The Farm that sells spores, this time those needed to grow fungi such as shiitake and other commercially grown exotic mushrooms. Mushroompeople does not produce the spores in-house, but distributes spawn purchased from suppliers to a nationwide customer base of commercial growers. Another element of this business is education, providing instructional materials to new farmers seeking to enter into commercial mushroom production, as well as individuals wanting to grow shiitake and other medicinal mushrooms for personal consumption.

Utilizing Technology

It almost goes without saying that the ability to change and adapt is paramount for survival in small business. By paying attention to the evolution of technology and its effect on business and the marketplace, those who choose to live in community can continue to keep pace with the rest of the world. Too much isolation and you will get left behind.

Unlike some segments of the counterculture, The Farm has in many ways always embraced technology, really from its very inception. Reel-to-reel audio recording machines were used to capture Stephen Gaskin's talks, providing the material for his first books. Rock and roll, with its plethora of amplifiers, flashing lights and other stage gear, became The Farm's calling card, using The Tennessee Farm Band as a marketing tool for attracting new members. Members of the band set up a recording studio in an army tent and began producing albums, simultaneously becoming one of the community's first small businesses, providing this service for musicians from outside the community as well.

Another member of the early Farm had a small movie camera. This was followed by the purchase of a reel-to-reel video recorder, documenting the community's arrival in Tennessee and The Farm Band's first tour. In 1978 the community purchased one of the first systems that recorded video to a cassette tape, and video documentation of all aspects of community life began in earnest.

Members of The Farm began experimenting with and using crude computers as early as 1980 for tracking inventory and development of a customer database. Using their freshly honed computer skills, in the mid-80s, two members assembled a database of Middle Tennessee voters and helped a Democratic congressman win an election, well before most people had even heard about computers.

Now, virtually every business in the world is completely dependent on computers for dozens of different aspects of their operation. The evolution of both the computer and the Internet has totally transformed the way people do business and simultaneously created many new types of businesses. This is definitely true on The Farm. Throughout the community, people sit in front of computer screens, tracking inventory, engaged in marketing, taking orders, bookkeeping, editing books, building websites, and the list goes on and on. A solid background in computers is one of the most important skills to have when determining how to earn a living in community.

Village Media

One of the best illustrations of the importance of staying abreast with technological advances would be to take a closer look at one of The Farm's high-tech companies, Village Media Services. It started in the

Early on in the community's development, The Farm embraced high-tech skills, leading to the formation of several small businesses. Above: Village Media specializes in video production, audio recording, website design and hosting, and other multimedia services.

early 1980s as the satellite installation company Nashville Satellite Electronics, which itself was born from the early experience in radio communications during the 1970s. It became evident early on that the work wasn't about satellites; it was distributing video throughout a home, to multiple rooms in a hotel or to the many homes in a subdivision. This led to contracts installing video distribution systems in corporate office buildings countrywide.

The owner of the satellite installation company began supplementing his income by writing for trade publications about satellite technology. He wrote the articles on an early consumer computer, a Commodore 64, saved them to a floppy disk and then sent them by Federal Express to the magazine publishers in various states. By the mid to late 1980s, the text of the articles could be sent by computer modem over a phone line.

In 1987 one of the magazine publishers changed its focus to a new emerging consumer technology, the home video camera/recorder or camcorder. To keep in step, the business became Total Video Publications, producing six or more articles each month about all aspects of home video.

The very first job doing video production came from The Farm Building Company. It had won a contract to reconstruct a 1800s log cabin in a Mississippi state park for the US Army Corps of Engineers. The contract specified that the entire project had to be documented in both photographs and video. After watching the raw footage, the Army Corps extended the contract to include editing the many hours of tape down to a 20-minute video. The money earned from this contract was used to purchase the professional video equipment needed to do the editing, thus making it possible to add video production services as another revenue stream.

Even in the small towns of rural Tennessee, awareness was growing for the potential of video. Local businesses could now afford their own television commercials and insert them into national networks, a service of the regional cable company. Manufacturers saw the benefits of using video to market their products and as a more efficient way to provide training for their employees. The wedding video became almost as essential as the wedding cake. Even the nearby Summertown high school began to assemble a video yearbook, featuring highlights of sporting events, skits and photo montages of the graduating students.

The fellow who had done the early recording of Stephen Gaskin back in San Francisco and later started the community's recording studio joined Village Media and brought along his audio engineering skills and a rack of gear to the mix. The company was now able to market itself as a small recording studio, bringing in songwriters, performers, local bands and even gospel groups to produce demos and albums.

About this time, the computer had transformed from a text screen to a graphic interface and a tool with virtually unlimited capability. Brochures, newsletters and catalogs could be produced on computers and became additional services supplied by the company. Video editing was revolutionized, leaving the analog realm of tape for the digital world of the computer. Audio made the same transition from cassette tape to CD. Instead of printing on paper, brochures, catalogs and even business cards could be reproduced on CD-ROM discs. Changing with the times, Total Video Publications moved on to a new name, Village Media Services.

The Internet changed everything. Every business on the planet needed a website, and Village Media was there to offer this service to companies in the nearby towns. The Internet also made it possible to communicate with people anywhere at virtually no cost. One video project produced for an international non-profit included collaboration over the Internet on a script, along with weekly online reviews of the edited material by representatives in Denmark, Germany and Australia.

It hasn't necessarily all been easy. One year, Village Media lost a third of its business overnight when the regional cable company took the production of commercials in-house.

The Internet also had a profound effect on magazine publishing. The writing of articles that had been a primary source of income for Village Media over 20 years completely disappeared when the magazine publishers went under because the bulk of advertisers abandoned print media and put the dollars into online banner ads and other forms of exposure on the Web.

The Internet also changed the nature of video production. Rather than produce a marketing video, most companies began to simply rely on their website, often built by the owner's nephew working his way through college. If clients still used video, instead of needing 1,000 DVDs, these same companies could put a single copy of a video on their website.

To survive, Village Media has had to be very diversified. In addition to video, audio, Web and print services, it has advised clients as marketing consultants on the appropriate use of technology. With a background in music and stage performance, Village Media has produced small music festivals hosted by The Farm Community. Experience as storytellers dovetailed into the marketing and management of organized conferences and retreats featuring PowerPoint presentations, live music and motivational workshops. The ability to change with the times and be flexible and open to new ideas has enabled the owners of this small company to maintain a lifestyle that preserves their independence and freedom in a way that fits their needs and quality of life.

Non-profits and Support Staff

The Farm got its first experience with non-profits when it formed Plenty International back in 1974 to facilitate the community's desire to develop international relief and development projects. The Farm Midwifery Center and The Farm School (The Farm Religious and Educational Center) were also set up as non-profits. After The Changeover, many of these organizations had to advance their own development in order to hire staff and provide an income for the people running the organizations. Today, although the salaries can be modest, the non-profit sector does make up a substantial percentage of jobs for people working inside The Farm Community, generating income in a variety of ways.

Plenty International employs a director, and his wife functions as the office manager and bookkeeper/accountant. Plenty also provides part-time employment for a few people in its office. Money for the organization comes from donations by individuals, businesses, foundations and grants. Grants to fund projects include a small portion or percentage earmarked for administration.

The Farm Midwives continue to deliver babies for women who come to The Farm to utilize their services. Most of these women are from Middle Tennessee or northern Alabama and drive to The Farm when their labor begins. A sizable number also come from across the country and around the world, living on The Farm during the last days or weeks of their pregnancy. The Midwives also manage an office for the North American Registry of Midwives (NARM), an organization co-founded by midwives from The Farm in alliance with other midwives throughout the US. NARM has developed and supervises certification for Certified Professional Midwives (CPM), a legal status now recognized in a growing number of states, thanks to its lobbying efforts. As in all small businesses, work related to both the workshop program and NARM revolves around office management. This means that the employment opportunities are not just for women with midwifery skills, but also for people with office and computer skills, including some of the Center's own midwifery apprentices.

Workshops and Retreats

The midwifery program is also the best working model in The Farm Community of a business based on workshops and retreats. The workshop business model succeeds for The Farm Midwifery Center because it offers women passionate about a subject a unique opportunity to spend time with instructors who have well-established reputations and experiences.

The Farm has developed several businesses based on conferences, workshops and retreats which attract visitors from around the world.

An Internal Economy

Because of its relatively large population base, when compared to many other intentional communities, The Farm is able to maintain an internal economy that supports a significant number of people. Some individuals, particularly those in the building trades, derive their entire income servicing the homes and businesses on The Farm. The Farm Store has some customers from outside the community, but the majority are residents. It was able to survive after The Changeover when most of the community was still struggling to generate an income, to a certain degree because the children of the founding members were still living at home, most just entering their teen years. After school each day, the kids would gather at The Store and purchase snacks. In the late '90s, when this group had grown up and were off at college or exploring the world at large, The Store's income dropped as well. The managers modified their business model by incorporating a food co-op that helped Farm residents save money while increasing the volume for The Store. It also put more emphasis on its lunchtime offerings, preparing homemade soups and salads.

Even a high-tech business like Village Media is able to benefit from The Farm's internal economy. While Village Media needs outside clients in order to derive a sufficient cash flow, it is able to also provide services for Farm businesses and non-profits, developing websites, marketing videos and supplying other communications needs.

Although our national economy is increasingly dependent on international trade and services, it remains to be seen how long this global trade system can be sustained. Maintaining a strong, locally based economy is an important element of survival. Developing skills that can satisfy the needs of your surrounding community provides some hedge against unemployment, market swings and inflation.

What Works

The ability to earn a living and support oneself and a family remains one of the biggest challenges for The Farm, especially for new people joining the community. Farm businesses generally find it takes about a year of training to integrate a new employee and are therefore reluctant to hire someone in the early stages of the membership process, at least until it becomes clear the person is likely to be around for a while. This means incoming prospective members must be prepared either to telecommute or find employment nearby.

Unlike smaller communities that can be very dependent on one specialized business, The Farm has many different types of employment, meaning it can better weather economic downturns and a struggling economy. In the end, each person must find what works for them, a way to generate income that does not compromise their ideals, but ideally fulfills their vision of a better world.

CHAPTER 3

Health Care and The Farm

FROM THE VERY BEGINNING, health care and The Farm Community have been intrinsically linked. One could boil down this relationship to its basic philosophy of simply taking care of each other. Throughout The Farm's history, this core concept and its implementation have revealed the complexities we all face when attempting to live up to our ideals.

Indeed, the US health care system almost took the community to the brink of financial collapse. The nature of health care on The Farm Community has gone through many stages of evolution, never perfect, but always adapting to the changing times. While no one is immune to the rising costs of health care and our dependence on the corporate medical system, in the end it comes back to where it all started, doing what the corporate medical establishment can never do, the personal care that only a family can truly provide.

Midwifery Opens the Doors

By the late 1960s, the hippies of San Francisco were already beginning to discover the responsibilities of adult life, as flower children transformed into parents themselves. The act of giving birth was but one more clash of cultures, as young women on the cusp of liberation found themselves strapped down, drugged and intimidated by the

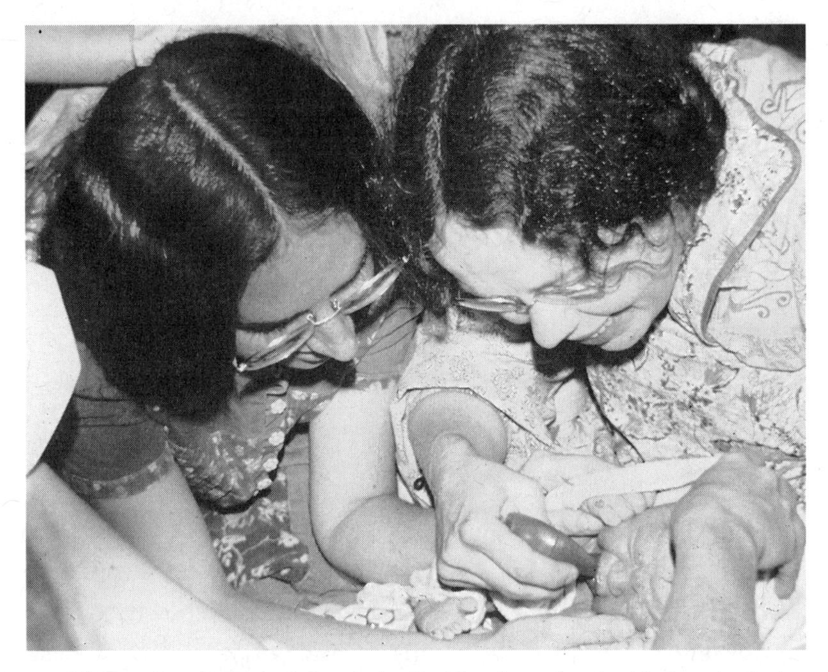

Health care is a human right, embodied by the role of midwives in their service to both mother and baby.

male-dominated medical establishment. New mothers emerged from hospital births sharing horror stories with their sisters in the movement. It was clearly understood that a natural, biological function had been transformed into a medical procedure that was oblivious to the seemingly obvious true miracle taking place.

As The Caravan departed from the West Coast on its epic journey, several women in the group were pregnant. There was no clear plan; it was more of a certainty that they would do as women have done for centuries and still do all around the world, bringing new life into being when the time was right, following the guidance of nature and life force energy.

And so it was that in a Wyoming parking lot, with a minimum of effort and a rosy glow, the first baby was delivered by the women who would one day become known as The Farm Midwives. Over the next several months, a total of ten babies were born on The Caravan, empowering and reinforcing the critical decision to take control of the future community's health care destiny. But there were difficult lessons to be learned. Not all births were storybook happy endings, with one baby, Ina May Gaskin's, dying after spending only 12 hours in this plane of consciousness. Along the way, a Rhode Island obstetrician came to visit The Caravan to share his knowledge and provide the aspiring Midwives with some basic instruction, such as how to identify signs of fetal distress and what to do if the baby came out with a cord wrapped around its neck.

Life and Death in Tennessee

When the San Francisco transplants finally settled in Tennessee, one of the first ways they sought to establish autonomy was to visit the local health department requesting birth certificate forms, to be filled in after each birth. The County responded favorably and with a strong dose of reality. Along with the birth certificates came a stack of death certificates, sending the clear message that the responsibility of care goes full circle.

The Farm's midwifery was aided in no small part due to the Tennessee state government's absence of regulations. The rural poor had a long history of home birth and the use of granny midwives. In the early '70s, there were no laws that regulated midwifery, which made it possible to establish the practice of lay midwives by this new generation of care providers.

It was also to The Farm's benefit that, in an adjacent county, a sizable old-order Amish community was flourishing. A precedent had been set for an independent religious group to embrace and rely on home birth as the accepted way for children to enter the world. Amish mothers gave birth assisted by immediate family members and, at times, with the attendance and support of a local country doctor. When The Farm came along, County officials already had a precedent established. The Farm Midwives were able to slip right in.

For many years, Dr. Williams had been one of the few area physicians to treat the Amish, even travelling out to their farms for house calls. Early on, he encountered members of The Farm and was intrigued and perhaps even a bit amused at their idealism and sense of purpose. As another example of the open and relatively unobtrusive nature of Tennessee government at the time, a physician was allowed to train anyone they wished to work under their license and authority. Dr. Williams agreed to provide training in basic health care to the Midwives on The Farm. A small clinic was opened inside The Farm, and he visited once a week, with Farm patients lined up for his diagnosis and treatment. In trade for the training, Dr. Williams asked The Farm to give back by taking on the medical needs of the Amish community, providing primary care and assisting in the delivery of their babies.

The Farm Clinic

One of the principal philosophies embraced by The Farm was the concept of Right Livelihood. Borrowed from Buddhism, Right Livelihood essentially means that a person's work should be seamless with their ideals. There could be no clearer path to Right Livelihood than providing care and promoting the health and well-being of others. Gaining knowledge and skill in health care became a cornerstone in the foundation of The Farm Community.

Initially the principal people involved with The Farm Clinic were the Midwives. During these early years, literally hundreds of babies were delivered, with the Midwives on call 24 hours a day. Prenatal care began as soon as a woman discovered she was expecting, with regular visits to the Clinic to monitor blood pressure, weight gain, diet and every aspect of the mother and baby's health.

It was obvious right from the start that care for mothers and babies was but one aspect of the bigger picture, that the broad spectrum of needs in a community of several hundred people was unlimited. Cuts needed to be stitched, infections treated and endless other aspects of the human condition addressed. The work quickly became more than this small group of Midwives could handle, and others began training. Roles began to become more specialized. A pharmacy was set up to dispense medicines. The Farm lab ran blood tests, grew throat cultures, checked stool samples and became an indispensable part of the equation. After a few years, two hippie doctors joined The Farm and became part of the team.

In the very beginning, the Clinic operated out of one of the rooms in The House, the original home of the former landowners and the only building on the property besides a couple of barns. In need of its own space, the community soon purchased a used mobile home that was set up with a waiting room and several examination rooms.

After a few years, the Clinic was moved to a real building. Farm work crews had learned how to pick up and move entire buildings and set them up in a new location. A church building from nearby Summertown was moved onto the property and set up in the center of The Farm's downtown area, providing easy access to most residents.

The First Crisis

Camped on land adjacent to the property the group would eventually buy, The Farm Community had to face its first serious health crisis, one that affected virtually every member and even the surrounding neighbors and local towns. In the spirit of the day, while exploring one of the many creeks flowing nearby, one person discovered water-cress, a delicious green leafy plant with a peppery flavor that grows in water. A quantity was picked, enough to provide a tasty salad for everyone.

What no one realized is that the clear streams of this middle-of-nowhere rural paradise could also harbor invisible, deadly diseases. Outhouses were still quite common among Tennessee residents in those days, with groundwater seepage from privies running directly into the creeks.

Days after the salad was consumed, a few people started showing symptoms. The culprit: hepatitis. In the end, over 100 people contracted the disease. Some became very sick and quite yellow, but fortunately no one died and many more avoided infection, the result of an important decision, one that challenged the community's spiritual philosophy and practice.

All hepatitis outbreaks must be reported to local health authorities, and in this regard, The Farm played by the rules. To prevent others from contracting hepatitis, those who might be exposed are given a shot, a vaccine with serum derived from horses. Months earlier, The Farm had made the decision to be completely vegan and had eliminated all animal products from the community diet, even honey. A decision had to be made. Would the community compromise on its principles and give everyone an inoculation?

Naturally this had local health officials concerned. There was some initial resistance to the idea, or at least there was a pause and not an immediate acceptance of the treatment. The matter was further complicated by the fact that the area where The Caravan had landed was the source of a spring that provided the water supply for the town of Mt. Pleasant. If The Farm refused to receive the vaccine on spiritual principles, there was real concern that the outbreak could spread.

Ultimately Stephen and others decided that it was more important to compromise and give everyone exposed the shot. This demonstrated to local authorities that The Farm could and would make rational and responsible decisions, especially when concerning the health and safety of the whole, which included not just members of The Farm, but also the greater Tennessee community.

Communication

In a way, work of the Midwives became the driving force behind many of the technological advances in the community. By this time, The Farm's land base had grown to over 1,000 acres, and residents were spread out over several square miles, nestled deep in the forest, along ridges and back hollers. Quick, clear communication became a challenge and a priority, especially if a baby started coming in the middle of the night.

One of the first ways to address this was a crude single-line phone system, which meant that anyone who picked up their handset could listen in or speak to anyone else on the line at the same time. Salvaged telephone line was strung up through the trees and run to the home of each midwife, as well as to public centers such as the food dispensary or Store, and the Clinic. If a woman's due date was approaching, a phone might be installed in her home, especially if she was expecting a second child and was known to have fast labor.

After a few years, the community installed an actual phone system, this time salvaged from a nearby phone company after they upgraded their equipment. With this new system, each home had its own phone (and number) that would ring when a caller used a rotary dial to place the call.

While this functioning phone system was a real step up, it still was inadequate in real emergencies. Like their counterparts in the outside world, Farm medical personnel began to utilize radio communication. Each midwife was supplied with a vehicle outfitted with a CB radio. Radios were also installed at other important communication centers, such as at the main entrance or Gate, where someone was on duty and listening in to the radio day or night.

EMTs and the Ambulance

One of the most direct ways to enter a career in medicine is to become an emergency medical technician or EMT. At first, several Farm members went to classes in a nearby town to get EMT licenses. It wasn't long before those who got the first training became instructors, holding classes on The Farm. Dozens of people signed up and earned licenses. Some continued their training and became paramedics, which allowed them to prescribe certain medicines and perform more advanced lifesaving procedures.

As the community's population grew, so did its medical needs. Medical emergencies are a basic fact of life any time a large number of people are assembled together. The bustling midwifery service meant that medical backup could be needed day or night. As a result, the community established its own ambulance service with teams of trained personnel on duty 24 hours a day. A small building was allocated to be the dispatch office and headquarters for the EMTs, with two fully equipped ambulances parked outside. Radio

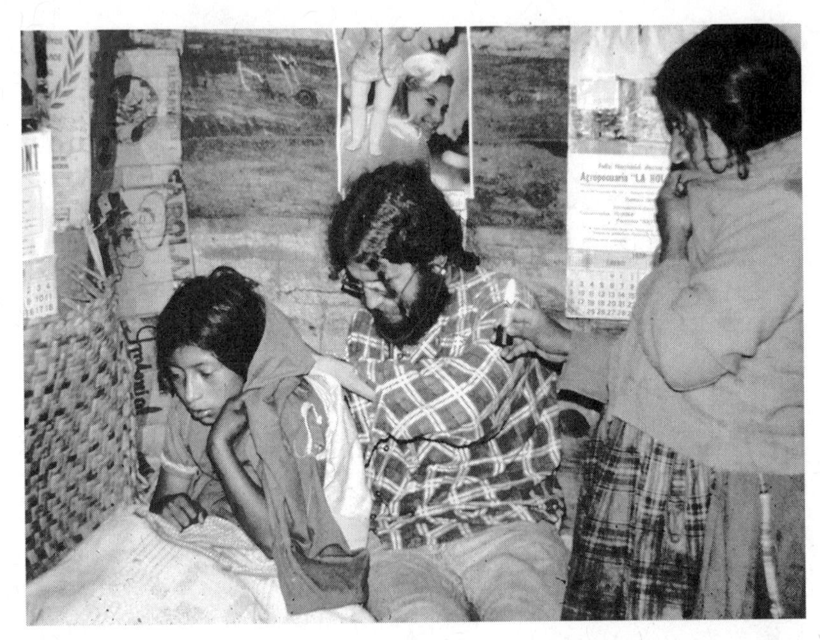

Volunteers from The Farm responding to an earthquake in Guatemala in 1976, soon found themselves administering medical care to the impoverished Mayan people.

communication linked the EMT Shed with the Gate, all of the Midwives and the community's two doctors, along with a separate radio system that could talk directly to the town hospital 30 miles away. A backroom was set up as the workspace for service technicians, whose job included installation and maintenance of the radios and all other electronics related to the Clinic, from calibration of the white blood cell counter to repairs on centrifuges and autoclaves.

Stepping Out into the Third World

In 1976 The Farm's newly formed relief agency, Plenty, responded to a devastating earthquake that had struck the poor Central American country of Guatemala. This launched a four-year adventure in which The Farm's newly acquired skills in health care played a central role.

The number of volunteers working at The Farm's camp in Guatemala swelled to over a hundred. Naturally, part of the "community kit" included medical personnel, giving the group the ability to treat minor injuries as well as have a line of defense against the more prevalent dangers such as parasites and infectious diseases like hepatitis and tuberculosis.

Word spread among the local populace about the Plenty team's medical skills, and before long, impoverished Guatemalans began arriving on the camp's doorstep seeking treatment and medicines. Responding to the obvious need, the team commandeered a vehicle to serve as an ambulance with runs dispatched day and night into the surrounding villages, saving lives and further expanding the reputation of Plenty.

Unfortunately the government did not appreciate a team of hippie volunteers doing more than the local health authorities to take care of the people. The situation became so tense that Plenty was forced to leave the town it had been working in for two years and relocate several hours deeper into the interior of the country; as part of this move, all medical services by Plenty ceased.

The South Bronx

Meanwhile back stateside, The Farm and Plenty discovered there were pockets of people living in third world conditions right in the US, notably the South Bronx. In many ways, it seemed to have been abandoned by the City of New York. Because of the amount of street crime and muggings, emergency responders and ambulances refused to go there.

In the winter of 1980, a national television news show broadcast a segment about a small group in the Bronx that had taken over an abandoned building. "Squatting" was the word used in those days, and this got wheels turning back on The Farm.

A call went out for volunteers, and a team was sent to locate a building in the Bronx that could house several families and become the base for a new project, a free ambulance service. A top-notch crew was assembled, veteran EMTs, paramedics, even a midwife. Carpenters and others with construction experience were also sent to provide a source of income that could support the effort.

The Bronx residents were a bit taken aback at first. Who were these white hippies plopping down in the middle of their scene? Ultimately the word spread, and before long, the project had two state-of-the-art ambulances running 24/7. In addition, the ambulance crew started conducting classes training Bronx residents to become licensed EMTs and fill shifts on the schedule. The respect of the Bronx community was earned.

"I remember one time we were the first one to an accident, and eventually the ambulance from the City arrived as well," describes James Mejia, a paramedic and one of the crew chiefs at the time.

> The patient was loaded into the city ambulance, and I followed along, continuing to administer CPR. My partner also jumped into the front with the driver and together went on the run to the hospital. Once we were at the hospital, we saw each other, and both of us exclaimed, "Oh my God, who is back with our ambulance?" The city guys were hysterical, certain that the unit had been stripped cleaned. We found a ride back and arrived to find our ambulance, lights flashing and all the doors wide open. Not a single item had been touched.

After about five years, Bronx residents trained by Plenty were employed by the City, doing ambulance runs into their home neighborhood. The response time for an emergency call was reduced to under five minutes, on par with service to other parts of the city. The Plenty Bronx Center was able to close its doors in 1985.

Epidemics on The Farm

The Farm's population continued to grow, but its infrastructure did not keep pace. Only about half of the homes had running water from the community water system. The rest were serviced by a truck that delivered water daily to fill tanks at each household, an inefficient system at best. There were no inside toilets, only outhouses. Each household was like its own small commune, ranging from 15 to 45 people, and very overcrowded. The close quarters and poor sanitation made it difficult to prevent the passing of germs, especially when there was a constant influx of new people unaccustomed to intense communal living.

The revolving door of visitors and new members, along with immigrants and international visitors, meant there was also constant exposure to new germs and infections. In the mid-70s, there was a second hepatitis outbreak. Those coming in from projects in developing countries also brought parasites and passed them along. Hand-to-mouth bacterial infections like giardia and shigella were commonplace and often difficult to eliminate.

The enrollment in the School of Hard Knocks meant Farm members learned the real-life definition of "commune-icable" diseases and how to prevent them. "Wash your hands" became a mantra; intensive cleaning and disinfecting, a holy duty. Through diligence and perseverance, these illnesses became less of an issue and gave folks experience and knowledge that would remain valuable for a lifetime.

Health Care and Financial Woes

On top of all this, The Farm was struggling financially, with its weekly income from work crews and businesses falling far short of its weekly expenses. Even more critical was the burgeoning debt, brought about in no small part due to the community's interaction with the corporate medical establishment.

The early decision to take personal responsibility for The Farm's medical needs did not factor in that at times it would be necessary to transport residents for emergency and long-term care in hospitals. In any small town or group of people, over time there will be a percentage that face some level of health crisis, be it from an accident or illness. The Farm was in no way immune to this.

One of the early shockwaves to the collective psyche came about when one of the men working in town for The Farm's tree trimming service fell over 50 feet and was severely injured, requiring several months in the hospital. Another man was servicing a piece of farming equipment when his pony tail got tangled in the mechanisms, ripping off his scalp. A horse kicked a young woman, shattering her knee. The list goes on and on.

In an effort to free itself from unjust systems, The Farm did not use medical insurance. In an ideal scenario, its own medical people would be able to handle most needs, with the collective economy able to function much as an insurance company does, pooling assets and then dispensing a small portion of those to cover the individuals needing treatments or procedures from a hospital in the city.

During the 1970s and early '80s, each adult member of The Farm had signed a vow of poverty and lived on an annual income of about $500 per year or under $2 a day, well below the poverty line. Although those requiring a hospital stay would have qualified for Medicare and government assistance, The Farm had a policy of not accepting welfare and so members did not apply for these benefits. Consequently, patients from The Farm were regarded by hospital accountants as members of a corporation, a corporation with significant assets, including The Farm's land base.

By the early '80s, the years of debt accumulation had reached a figure of over half a million dollars. Pressure was mounting, and there was some indication that one of the larger hospitals could initiate a lawsuit and seize The Farm's land in order to satisfy its debt. Finally one did, placing a lien against the land. It is clearly evident that medical bills were a major factor in the community's restructuring of its economic system in September 1983.

The Changeover and Community Health Care

While some aspects of The Farm's collective economy resulted in dynamic achievements, in numerous other ways and for many years, it had been a failure. The years of voluntary poverty and without adequate funds for medical care meant simple procedures like a hernia operation went untreated. Living conditions made it difficult for those who contracted a serious illness to heal, and they were forced by circumstances to leave. Even getting a new pair of glasses required approval by a central authority under pressure to fairly distribute extremely limited funds. It was definitely time for a change.

The basic concept for the new economic system and its relationship to health care was pretty straightforward. On the post-Changeover Farm, each member and family would be responsible for its own medical expenses. The Farm would no longer pay for visits to doctors in town, surgeries or any other medical or dental care outside the community.

However, one part of The Farm's health care system did survive The Changeover. The community agreed to continue operating its own internal clinic, with one person to receive a small salary as a primary care provider for the community. Although The Farm no longer had its own doctors (both had left prior to The Changeover), The Farm Clinic was able to maintain a relationship with a local doctor and work under his authority.

This meant that a portion of monthly membership dues went toward the expense of operating the Clinic. The community was able to continue providing basic care, checking throats, ears, rashes, colds and a long list of ailments that are a part of everyday life, especially when raising children. These dues also paid for all overhead, including electricity, gas for heat and hot water, and maintenance of the building.

On the other hand, The Farm could no longer afford to staff and pay for its own 24-hour ambulance service. Instead, those who had been in this role agreed to continue functioning as volunteer first responders, on call should an emergency arise. An ambulance was parked and kept ready to leave on a moment's notice.

One day, while on an emergency run to the hospital, The Farm's ambulance was involved in a serious accident. All of the people were uninjured, but the vehicle was totaled. The legal qualifications for ambulance vehicles are constantly upgraded, meaning that the purchase of a new(er) ambulance was out of reach. By this time, The Farm's population was now too small to economically support this level of technology. Life was much different than it had been on the old Farm.

Now most families had one or two cars and could drive themselves to the hospital if the need arose.

Throughout the '70s and '80s, dozens of people had received informal medical training through The Farm's internal apprentice system. Now faced with the need to generate an income for their families, many of them went back to school in order to gain actual licenses and degrees. The Farm was fortunate that a nearby community college offered a two-year program in nursing. In those first years after The Changeover, nearly two dozen women graduated as registered nurses, most with honors or at the top of the class. Some continued their education, becoming nurse practitioners, certified nurse midwives and physician's assistants.

Like everyone else, The Farm Midwives had to face the reality of generating an income. They could no longer offer their services for free to everyone who came through the Gate. The Farm Midwifery Center was set up as a non-profit, and the Midwives began to charge a fee for their services. The decision about what to charge was difficult. All of the Midwives still believed that delivering a child into the world was a spiritual calling and an act of love. Often those seeking natural childbirth were either members of The Farm or like-minded souls from the greater counterculture in the US who lived a subsistence lifestyle and had limited funds. Still it was only fair that the Midwives be compensated for their efforts, and the fees they asked for were still well below the cost of a hospital birth. Even the Amish clients were expected to pay a token amount.

In the early '90s, members of The Farm decided through a democratic vote that the community no longer needed to pay someone to be available as an on-call care provider. There were enough skilled medical people scattered throughout the community to handle minor illnesses, and if anything was considered serious, most folks simply made an appointment with a doctor in town. The person who had been salaried by The Farm continued to provide care to anyone requesting her services, but charged a small fee.

Health Care on the New Farm

As more people in the community began to develop skills and get on their feet economically, they have begun to purchase health insurance. The two largest and most successful Farm businesses provide health care coverage for their employees. Others may receive insurance from their employment outside the community. Still, because each family's finances are different, some cannot afford insurance or choose to rely on luck and a healthy lifestyle.

In many ways, The Farm's lifestyle choices have had a direct impact on the health of its residents. Established initially on a vegan diet (no dairy or eggs), most members remained vegan or vegetarian after The Changeover. This has meant that, for the most part, Farm members do not struggle with the weight gain and obesity so prevalent in this country today. Another founding agreement was that there would be no tobacco, and its use continues to be very limited. In general, people maintain an active lifestyle, by walking, running, biking, hiking and a host of other physical activities. Problems like heart disease are virtually nonexistent, with the rare exceptions of individuals who may have a history of heart disease in their family.

A Network of Support

Ultimately, members of The Farm have had to recognize that they are not separate from society as a whole and from the struggles all humanity faces as we deal with disease, illness and the unpredictable.

And so in a way, once more it comes back around to doing what we can, when we can. No one on The Farm is alone. When a person or a family is in crisis, they have close friends who rally with support, supplying food and rides to medical appointments, running in shifts to remain at their bedside, whatever it takes, to each as they have need.

In the event of an emergency, within minutes, there can be 20 trained medical professionals on hand, checking blood pressure, listening to a heart, powering up the defibrillator. One Farm member oversees health care for our region in the state, another worked as a primary care provider in the local doctor's office. Farm nurses have worked in the nursery of a local hospital, been in charge of the pediatric floor of another, and the list goes on. Two MD's have homes in the community.

And The Farm's connection to the medical community is not just in Tennessee. The Farm is a tribe, a network of like-minded people spread out across the country and around the world, people who share common values that have been personally touched in some way by this unique social and cultural experiment. Following the spiritual path of Right Livelihood has led so many to careers as nurses, doctors, homeopaths, physical therapists and hands-on healing modalities of all different stripes. When a Farm or former Farm family or member encounters a medical crisis, they can take advantage of a vast support

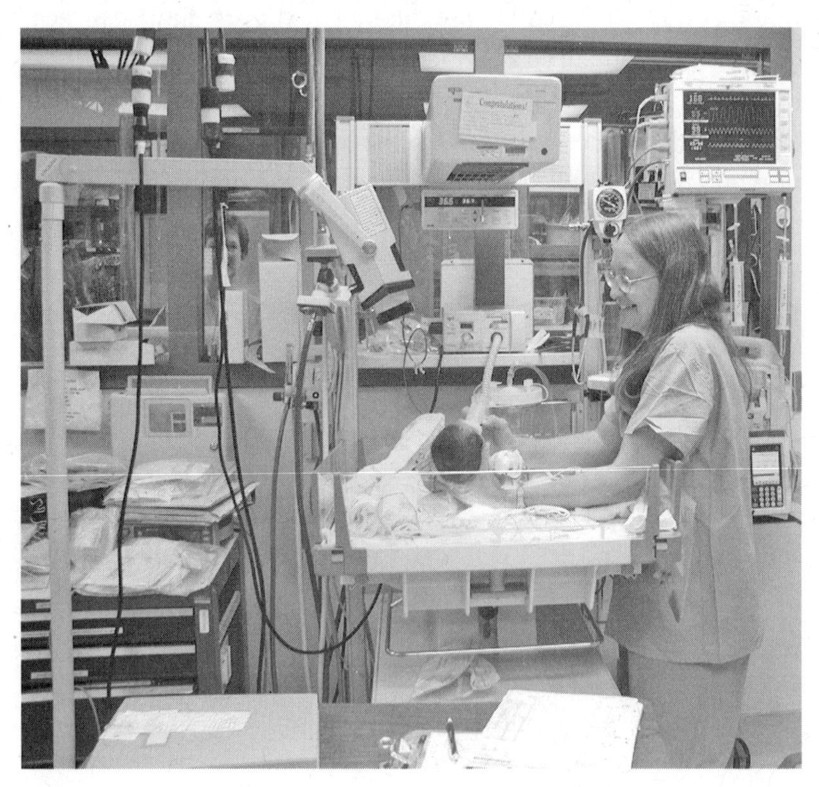

Taking care of each other.

network, from the friend whose son is now a naturopath doctor in a cutting-edge cancer treatment center to the highest levels at the Centers for Disease Control.

The Full Circle

Perhaps the hardest lessons are when someone loses their battle and crosses over from this life into the unknown. With founding members of The Farm reaching their 50s, 60s and 70s, losing a loved one is becoming much too common. Over these many decades, The Farm's cemetery has grown and expanded. Those who have gone before have demonstrated such incredible courage, grace and dignity. They have been role models and brought to the community what some call a new side of midwifery, the skill of hospice. A time to live, a time to die.

Taking care of each other. Like it says in the vows. For better or for worse. In sickness and in health.

At times, it seems life consists primarily of overcoming trials and threats to our well-being, so that we can get back to the task of finding happiness and contentment. Ideally, we succeed in overcoming these difficulties unscarred and with little setback, but that is not always the case. Entropy prevails, but for a time, we are able to overcome and push forward, and with grace and gratitude, we give thanks to those who have dedicated this, their lifetime, to gain knowledge and skill in the healing arts.

CHAPTER 4

Health and Diet

Over the decades, The Farm's attitude toward diet has gone through many transitions, from completely random to strict conformity, from rebellious to practical. In the end, it has returned to a place of personal choice and practice, ultimately defined more by general agreement than rules.

The Evolution

In the early days of San Francisco and Monday Night Class, discussions about diet were not really part of the picture. However, as the exploration of consciousness began to recognize the sentient nature of other life-forms, the nature of killing for food began to weigh upon the minds of developing pacifists, unable to reconcile the taking of life, be it through war or the systemized slaughter of other species.

When The Caravan left San Francisco, it departed without any defined agreements beyond a professed commitment to follow truth and a path of nonviolence. Each person was following their own spiritual journey. Some followed the clichéd "see-food diet: I see food...I eat it," consuming whatever was convenient, easy and cheap. Others had developed strict codes of permissible foods, garnered from a mix of medical research and scientific facts, gut feelings and hyped superstition.

Agreements surrounding The Farm's diet fell into place soon after settling on the land in Tennessee.

Arriving on the land in Tennessee, the group morphed from a random collection of spiritual seekers to an established "church" agreeing to live communally, collectively sharing fortunes and the development of a unified vision. How this would be defined in relation to diet was very often simply announced by Stephen, whose role was to articulate the shared agreements. The charted course for the nature of food on The Farm was always a mix of spiritual values and practical concerns.

In the quest to build a new society, going back to the land was very much about growing your own food. It was understood that there could be no better way to be in tune with the Earth and the environment than through this very direct relationship between humans and plants. Farming and growing food was seen as the purest form of right livelihood, work that was seamless with one's ideals.

This was also the first step in taking personal responsibility. For the hundreds of young people whose souls had become profoundly sensitized through psychedelic experiences, taking the life of another of God's creatures was a step most were unwilling to take.

In the back-to-the-basics belief system of the early Farm, in some ways, the fundamentalism of the community resonated with certain aspects of the Catholic Church with its dedicated respect of life force. In addition to the abstinence from the use of chemical or synthetic birth control on The Farm, the commitment to life force was extended to include the agreement that no animals would be slaughtered to feed the community. Even the consumption of honey was forbidden for a time, since this involved dominating and controlling another species.

Of course, there was a practical side as well. To raise enough cows to provide the hundreds of people living in community with beef would require vast amounts of land. In addition to a huge acreage dedicated to pasture, more would be required to grow crops such as corn, soybeans and hay to feed the cattle through winter months. It was recognized from the start that the community simply did not own enough farm land to accomplish this.

Awakened to the bigger picture, the community could not ignore a growing awareness of famine and starvation around the world and its relationship to the overconsumption of resources by the Western countries. Transcending individualism to the understanding that we are all one meant that there was a direct connection between a starving child in Africa and the obesity of America.

The science was clear: It takes ten times the amount of land to raise soybeans, feed them to cows and then eat the cows as it does to simply consume the soybeans directly and ingest the same amount of protein. If a person eliminates meat from their diet, the protein that is essential for basic nutrition must be replaced by a plant source, and soybeans contain the highest amount of complete protein of any legume. When combined with a grain, the amount of complete protein that can be derived from soybeans increases considerably. World hunger could be eliminated if the vast amount of resources consumed to raise cows and other livestock was diverted toward beans, fruits, vegetables and grains fed directly to people. A vegetarian diet was recognized as a revolutionary act and a political statement.

Milk and Dairy

When making the decision to eliminate meat from the diet for spiritual reasons, it also becomes necessary to consider the production and consumption of milk and other dairy products. Like all mammals, cows make milk to feed their babies. Humans are the only species that drinks another mammal's milk. Dairy farmers impregnate a cow, and when the calf is born, it is taken away and either slaughtered immediately and turned into veal or fed food supplements until it is old enough to eat grass and grains, and either eventually able to produce milk or butchered for beef. It only takes one bull to impregnate many cows, and most commercial farmers use artificial insemination. Since male calves will never produce milk, there is no question they eventually become part of the meat food cycle. The milk meant for the calf becomes the milk and various different kinds of dairy products sold in stores. Ultimately, it is clear that dairy is intertwined with the production of meat and cannot be truly considered separately.

A collection of terms has evolved to identify various distinct diets. People who no longer eat meat but still consume dairy products are more accurately called lacto-vegetarians, with the definition of "lacto" meaning "milk from a cow." Ovo-vegetarians eat eggs. Strict vegans will eat neither eggs nor dairy and also avoid leather, a by-product of the meat industry. The strictest vegans also consider honey an animal product.

In an effort to be conscious and consistent in spiritual and dietary practice, very soon after arriving in Tennessee, The Farm officially became a vegan community. No one wore leather, ate butter or cheese. Eggs were out. For the hundreds who had grown up on a typical American diet, it meant learning and developing a whole new way to cook.

Countering Superstition

Originally coined to describe the innocence of the counterculture, on The Farm the term "flower child-y" came to reference a naive belief system based on unsubstantiated dogma and superstition, often adhered to even in the face of obvious evidence to the contrary. With The Farm serving as a magnet attracting people from across the broad spectrum of hippie culture, the community was exposed firsthand to dietary practices of various extremes. One of the best (or worst) examples were people who adopted a "fruitarian" diet, eating nothing but fruit until the lack of protein caused their hair to turn prematurely grey and then fall out. The high sugar content of an all-fruit diet would also rot their teeth. Somehow these people were able to look in a mirror and overlook the obvious.

In the 1960s and '70s, health-food purists believed that white sugar was evil, causing all sorts of maladies. While there may be growing evidence to support these beliefs and the effects of sugar on the body, the desire for sweetener and its appeal to humans cannot and, in most cases, will not be denied.

Honey was expensive, and there was no way the budget of the early Farm could afford enough honey to supply the entire community. As mentioned earlier, honey and the task of raising bees had been decided against for philosophical reasons as well. Although Tennessee was regarded as the South, it was too far north to grow sugar cane. There was an alternative, a plant-based sweetener related to sugar that was commonly grown and processed by Tennessee farmers, sorghum molasses.

The Sorghum Mill

Like sugar, sorghum is derived from a cane and is considered a grass in the same family as corn. The cane is crushed and squeezed, extracting the juice. Only 16 to 18 gallons of cane juice are needed to make one gallon of sorghum molasses, compared to maple syrup, which requires as much as 60 gallons of sap to produce one gallon. This thick brown molasses has long been a staple in the South, and while it is widely known for its popularity on pancakes and hot biscuits, sorghum has also been recognized for its minerals, including iron.

Ecstatic over its discovery of a natural sweetener that they could grow, The Farm went all out for sorghum in a big way. One of the

Sorghum molasses is made from a cane that can be grown in Tennessee. It was used as a sweetener in the early years of the community.

first buildings to be constructed was a sorghum mill based on plans designed by the US government for commercial production of molasses.

Harvesting sorghum canes requires a lot of hands and physical labor. The leaves are stripped or removed from each cane using a machete. Another swing of the machete cuts the cane off at the base, low down to the ground. Usually the worker will cut off several canes before toting them over to the truck that will haul the load back to the mill.

There the load of canes was dumped in a big pile at the top of the hill above the mill. It was one person's job to shove the canes into or between two massive steel rollers of the "gin," crushing the stalks, squeezing out the juice, which flowed by pipe downhill to the mill. Old timers used a horse or mule to turn the rollers, but The Farm's setup was a little more modern, using a PTO (Power Take Off) wheel on a tractor to spin a large fan belt to rotate the gin's rollers.

The Sorghum Mill building had three levels, stair-stepping down the hill. The cane juice flowed into evaporators, metal troughs cascading in series from one end of the building to the other. Propane gas fires under the evaporators heated the cane juice, boiling away the water as steam simultaneously concentrated the remaining sugar syrup. Impurities formed as a green scum on the top of the syrup, which was removed by teams of "skimmers," people armed with screened strainers that would lift off the scum while leaving the thinner syrup behind.

After the cane juice traveled down the divided channels of the evaporator, it reached the end of the line and the syrup would be sufficiently concentrated then siphoned off into quart jars. At first, since the Farming Crew was organized and run primarily by the men, the cooking of the sorghum was managed by men as well. However, after a few runs, it was soon discovered that the patient cooking skills of women greatly improved the quality of the syrup, which led to the women being put in charge of the evaporation process.

The idea of hippies making sorghum was fascinating to the local Tennesseans, and when word got around, people began showing up in droves wanting to purchase the jars to take home. A label was created, and "Old Beatnik Sorghum" became one of The Farm's first cash crops and products.

Unfortunately, as an everyday sweetener, sorghum left a lot to be desired. Sorghum has a strong, unique flavor, overpowering any other flavors in the mix. Even chocolate cake became sorghum cake. Any and all hot teas tasted more like diluted syrup. Sorghum did not mix well with cold soy milk, simply sinking to the bottom. There was also one other undesirable side effect: Consuming more than a taste on an occasional stack of pancakes or oatmeal gave people the runs.

Nutrition—Enter UNICEF

To a large degree, the direction of health and diet on The Farm was driven by the Midwives and the women who took on the responsibility of health care for the community. To grow a healthy baby requires a healthy mother. A healthy mother needs proper, balanced nutrition. The Farm began a serious study of nutrition in order to develop a diet that could be sustained and emulated.

As word about The Farm began to filter out through society at large, it caught the attention of members of UNICEF interested in dietary studies. The Farm represented a relatively large, defined group of people all eating the same, specifically a vegan diet. Nutritionists were dispatched to The Farm to run tests, examining the community's daily intake of calories, protein and other nutrients. It reached the conclusion that, by replacing meat with soy, the daily protein requirements were being met. However, where the general Farm diet fell short was in the consumption of carbohydrates. Because dairy products were eliminated from the vegan diet, cheese and butter were out, common sources of fat for many Americans. The absence of white sugar meant Farm members did not have extra calories to burn when actively engaged in the arduous work of building the community. In other words, people were skinny.

At the recommendation of the nutrition experts, white sugar was added to the Farm's weekly grocery purchases. Sugar did not violate the vegan ethic, since it was produced from either sugar cane or sugar beets. Margarine was also incorporated as a way to provide more fat in the diet on a daily basis. The amount of cooking oil per household was also increased.

With the health of the community's diet receiving official approval, there was less debate about what to eat and more energy focused on how to prepare vegan meals that were tasty and delicious. Soy became the staple. Standard recipes were converted and re-engineered with vegan ingredients. Substitutes for cheese and meat dishes were developed. After a few years, these recipes were collected and published in *The Farm Vegetarian Cookbook*, one of the first publications to provide easy-to-follow, good-tasting vegan recipes, and it became a commercial success for The Farm's publishing company.

The Farm's vegan diet was one of its firmly established agreements; one might even say a rule, something that every member of The Farm was obliged to follow. Since all food purchases were conducted collectively through a centralized system of bulk buys distributed weekly by The Farm Store, it was easy to maintain this standard. Everyone ate much the same because they had all the same ingredients and worked from the same recipes.

In an effort to be self-sustainable, the community depended on the Farming Crew to supply as much as possible. Because the fruits and vegetables were seasonable, at specific times of the year, certain foods were available in abundance and then absent when the season ended.

The canning and freezing operation, located at the center of the community, was a buzz of activity throughout the summer months. With so many mouths to feed, it seemed impossible to put up and store the amount of fruits and vegetables necessary to feed everyone through the winter, especially as the community's population grew to over 1,000 people. During winter months, when it is especially important to get vitamin C and other nutrients to stay healthy and fight off colds, their diet could be disappointingly bleak. As The Farm entered the 1980s, this was one element of dissatisfaction on everyone's mind.

Tempeh

Tempeh is a food that has its origins in Indonesia. Soaked, cracked soy beans were inoculated with naturally occurring mold spores by wrapping the beans in a banana leaf and leaving them for the mycelium to grow, turning them into a white mass. Seeking additional recipes for soy, members of The Farm learned of tempeh in their research and were given spore samples by scientists working with the USDA.

A few batches of tempeh were produced and samples passed around The Farm for a taste test. Deeming the new food delicious, The Farm team figured out how to generate additional tempeh spores by growing them on sterilized pieces of sweet potatoes inoculated in test tubes. Within a few years, tempeh had become a staple of The Farm vegan diet.

The story of tempeh, along with a few recipes, was published in *The Farm Vegetarian Cookbook.* As demand grew for tempeh around the country, there was also a need for spores. The Farm set up a commercial business for the production of tempeh spores, quite possibly the first in the world. It became a small but steady cottage industry for the community that continues to this day.

Nutritional Yeast

No story of The Farm's diet would be complete without mentioning nutritional yeast. Brewer's yeast was a popular health-food supplement in decades past because of its abundant B vitamins. However, brewer's yeast is a by-product of beer, which gives it a bitter flavor from the hops. In the early days, The Farm's nutritionists discovered nutritional yeast, grown on molasses so it tastes sweet. A staple of The Farm diet since the '70s, it is the principal ingredient for a cheese substitute on pizza, as a butter substitute on foods like popcorn and potatoes and sprinkled on beans, rice and vegetables. Most Farm homes keep a container of "nut" yeast on the table just like salt. Nutritional yeast is now commonly available in natural food stores.

After The Changeover: Freedom of Choice

The Changeover of 1983 had a radical effect on The Farm's dietary practices. Virtually overnight, food purchases were no longer controlled by a central authority. Everyone was on their own to purchase food for their family with their own income. The Farm Store became an actual store, with prices on everything. It continued to stock staples but also began adding a much broader variety of items.

For the most part, families went to town and shopped in local supermarkets where prices were generally lower than The Farm Store. After The Changeover, they were free to purchase orange juice, vegetables out of season, even previously forbidden foods like dairy and eggs. Not everyone was able to easily digest soy, and before

The Changeover, they were expected to just eat it and not complain. Now they could have other options. It became up to each family to determine what food they liked, could afford and worked for them.

In many ways, it was liberating. There was an unexpected effect in the economy of scale. For example, to purchase orange juice for the entire community for a week would represent an investment of perhaps a thousand dollars, maybe even several thousand dollars, an impossible expense when there were budget shortfalls every week. In contrast, for one family, the purchase of a few cans, a carton or a bottle of juice might represent only a few dollars, an amount virtually any family that had someone earning money could afford.

The End of Large-Scale Farming

Because every family was responsible for their own income and expenses, earning money became a priority. Financial losses that had been incurred by The Farming Crew's commercial ventures had demonstrated the difficulty in earning a living as farmers. With the community forced to eliminate any non-essential services, there was no longer a Farming Crew supplying food for everyone. A few families began developing home gardens, but the large fields that had been full of produce were left untended and abandoned. Most folks simply did not have the time or energy to work earning money and grow their own food too.

Fast forward 30 years after The Changeover, and the number of home gardens has risen significantly. Several acres of fenced community garden plots in The Farm's big fields give members access to plots with full sun and space for crops like sweet potatoes and winter squash, which require a lot of room. Members garden because they desire the better flavor of organic homegrown veggies and for the joy and satisfaction it brings.

After The Changeover, to preserve the integrity of The Farm's fields so that someday they could be used for farming again, they were converted into a hay crop. This ensured that they would get mowed a few times each year, which would prevent them from becoming overgrown with trees and returning to forest. Since the community was no longer farming, it did not have the equipment needed for a hay operation and began working with local farmers in trade. They would receive most of the hay in exchange for mowing the fields, with the community keeping a portion for those folks who kept horses.

Over ensuing years, The Farm eventually purchased its own tractors and mowing (but not haying) equipment and became less dependent on others to maintain its fields. Of course, that costs time and money. Once again, to cut expenses and ensure that the fields are cut and maintained, various local farmers have made agreements to cut hay on The Farm's fields.

Dietary Trends—
Convenience and Personal Choice

The Farm's dietary choices follow a bell curve, from a few who follow no pattern and eat anything they want, including meat, to the majority in the middle who are lacto-ovo-vegetarians, and to the hard core vegans who read all labels, follow strict personal guidelines and choose only the healthiest options.

People engaged in careers often find it necessary to rely more on foods of convenience. It has always been more difficult to maintain a vegan diet outside the community, especially when traveling. A person might choose to eat cheese as a way to get protein on the road. And besides...it tastes good. Others have felt the same way about eggs, seeing them as a healthy and convenient source of protein. For the most part, few of the original members have been interested in adding meat back into their diet.

Through casual observation and a longer perspective, it is also possible to recognize that diets follow trends and fads. One year, many people may explore gluten-free cuisine; another year, a new book pronounces the benefits of eating for your blood type, and a percentage of health-conscious Farm residents will follow its advice. Various diets are either adopted or abandoned as individuals find what works or what they simply have time for. Sometimes getting picky is simply more trouble than it's worth. Other changes stick and become incorporated by individual choice.

Our dietary needs also change as we age. A person in their 20s and 30s performing heavy labor and working at jobs that use a lot of calories requires meals with lots of carbohydrates. Meals served by the Community Kitchen of the old Farm would include generous portions of brown rice or other grains, breads, pasta and sugary desserts. Over the years, the community's dominant population of aging baby boomers began to desire meals tilted more heavily toward vegetables and salads. Farm cooks have had to modify their menus with changing times.

Another influence on The Farm's diet has come from a growing population of young people, children of the founding generation, or new residents who are of a similar age, influenced more by mainstream culture. A portion of the second generation has spent time outside the community growing up with a parent no longer a member. Others have gone away to college or to live on their own and explored new foods, including meat. They may move back to the community with a partner that is not vegan or vegetarian. The dietary choices of the next generation can reflect a different set of personal choices that may include meat or not. As their numbers grow, it becomes impossible to identify The Farm as adhering to any specific diet.

Immediately after The Changeover, food sold through The Farm Store continued to be vegan, but over time, its product line has grown to include dairy products and eggs, responding to the purchasing profile of its customers. No meat products are sold, reflecting as much the dietary choices of The Farm Store's managers as it does the broader population of the community.

Vegan, vegetarian or not has become a matter of personal choice. However, the general unspoken agreement remains that all community meals, even potlucks and snacks brought to meetings, would remain vegetarian and that food should be clearly marked if not vegan.

Soy: Good or Bad?

Over much of its history, The Farm relied largely on soy as its primary source of protein. However, in recent years, a number of studies have questioned the benefits of a diet based on soy, linking it with cancer and a myriad other issues. While it is important not to ignore any new information, it can be difficult to know how much of the anti-soy conversation is being driven by the meat and milk industry behind the scenes. There has been a concerted effort by industry lobbyists and media campaigns to combat the trends toward healthier lifestyles and away from animal-based food products.

In general, The Farm diet is still very pro-soy with many of its first, second and third generations living happy and healthy lives, all with a lot of soy in their diet. While our community has been around long enough to experience friends and loved ones who have developed or succumbed to cancer, it is impossible to attribute a link to soy when everyone in the world today is inundated with exposure to a seemingly endless list of potentially harmful chemicals, radioactive fallout and other possible environmental sources.

That said, some members decided to remove most "processed" soy from their diet, such as the fake burgers and sausages. It is also now

The Farm was a pioneer in the use of soy as a source of protein, producing its own milk and tofu in one of the first soy "dairies" in the US.

much easier to incorporate protein from other beans and nuts into a daily diet to achieve balance so that it is not overly dependent on one form of protein. Soy researchers often speak in favor of fermented soy as being the best option, foods such as miso and tempeh.

Many soy critics also point out that Asian populations do not have the heart and other health issues that plague the West, but utilize fish as their primary source of protein and consume soy as a secondary food source and not a staple. While it may be true that the protein and other nutrients from fish can be incorporated into a healthy diet, it is also important to consider what is sustainable for our planet. Wild fish populations are being rapidly depleted. Pollutants like mercury and other toxins are concentrated in fish. Farmed fish may not have the same beneficial nutrients as their wild counterparts. The question must be raised: Can fish be a sustainable solution for supplying our planet with protein?

Much of the world is still starved for protein, resulting in malnourished children with stunted growth. Soy continues to be one of the best solutions for addressing this problem.

Vegetarians must replace the animal protein they remove from their diets with protein from other sources. Soy beans have significantly more complete protein than any other plant source. It would be safe to say that The Farm as a whole still believes in the power of soy. However, it plays only one part in the complete requirements for a healthy lifestyle.

The Case for Eggs

During The Farm's communal period when vegan diet was the rule, no one ate eggs. After The Changeover, as many people moved more toward a less strict vegetarian lifestyle, eggs gradually grew in acceptance. They seemed especially beneficial for children, who are often picky about what they eat and may not consume enough other forms of protein such as beans or soy products to meet their daily requirements. Kids will usually eat eggs without a problem, and sometimes parents are relieved when they can provide such nutritious food. One egg contains 6 grams of high-quality protein and all 9 essential amino acids.

Back in the '80s, eggs received a bad rap because of their high cholesterol content. It is true that eggs contain 210 mg of cholesterol when the daily maximum is considered to be 300 mg, but nutritionists now distinguish between good and bad types of cholesterol, with eggs containing the good HDL cholesterol the body needs. They also contain many vitamins, minerals and special nutrients such as lecithin, important for metabolizing fats in the body.

Unlike dairy, which requires animal breeding and is indirectly tied to the meat industry, it is not necessary to kill a chicken in order to get its eggs. Eggs are actually the chicken's menstrual cycle, delivered in a neat and compact package. When a chicken is in its laying phase, it produces eggs on a regular, even daily, basis and they only can turn into baby chickens if they have been fertilized by a rooster.

After a number of years, a chicken will go through "henopause" and no longer will lay eggs. Compassionate owners will put these older hens out to pasture, so to speak, keeping them as part of the flock; they are much too old and tough to be eaten anyway. Some will also keep a rooster so that the eggs can be fertilized. Periodically they will allow a chicken to continue sitting on its eggs in order to produce a new flock of chicks that will replace the older non-laying hens. The rooster also plays an important role in the pecking order, monitoring and protecting the flock from danger, warning of impending danger, be it a wandering dog, a fox or a swooping chicken hawk.

For many years, folks on The Farm who wanted eggs could purchase them from the local Amish, knowing they were free-range and fed a healthy diet, even if the eggs technically were not certified organic. Eventually a few families started raising their own chickens as a way to get organic eggs, save money and become more self-sufficient.

Living in the woods with chickens does present its challenges. In addition to the occasional dog, the presence of a large number of predators means that the home chicken coop must be fenced top and bottom. Chicken wire fencing alone is not enough. During the night, raccoons will wander up from the creeks and will actually chew their way through the relatively light-gauge wire fencing. Each night, the chickens must be rounded up back inside an enclosed coop, but even that does not guarantee their protection. Snakes, weasels, possums and other critters find their way inside, consuming eggs, chicks and full-grown chickens.

When it is possible to let chickens wander loose around the yard, some Farm families have discovered another major side benefit. An expanding deer population within the community has led to a proliferation of ticks, not just a nuisance but a serious health problem. Chickens spend their day eating bugs and can keep a yard clean of these nasty critters.

Oils and Fats

As mentioned earlier, a study of The Farm's diet and calorie intake in the early '70s indicated the need for more fat and carbohydrates, resulting in the addition of margarine to the weekly grocery list. By the '80s, there was a growing awareness of the health risks of hydrogenated vegetable oils, the main ingredient in most margarine. Since then, there has been a great deal more data regarding the intake of oils and fats. As might be expected, some natural health experts go to the other extreme, advocating the elimination of virtually all fat from the daily diet.

The good news is that the increased awareness about fats and oils has resulted in a much broader selection of different oils to choose from. Once again, moderation combined with intelligent and proper use allows us to enjoy the benefits of fats and oils when developing a healthy diet plan.

Diet has so much to do with culture, personal preferences and identity. Ultimately, it is possible to be a healthy person with a wide range of dietary choices. In the end, it is important to remember not to judge someone by what goes into their mouth, but pay attention to what comes out of their mouth in the form of words, ideas and the expression of their values.

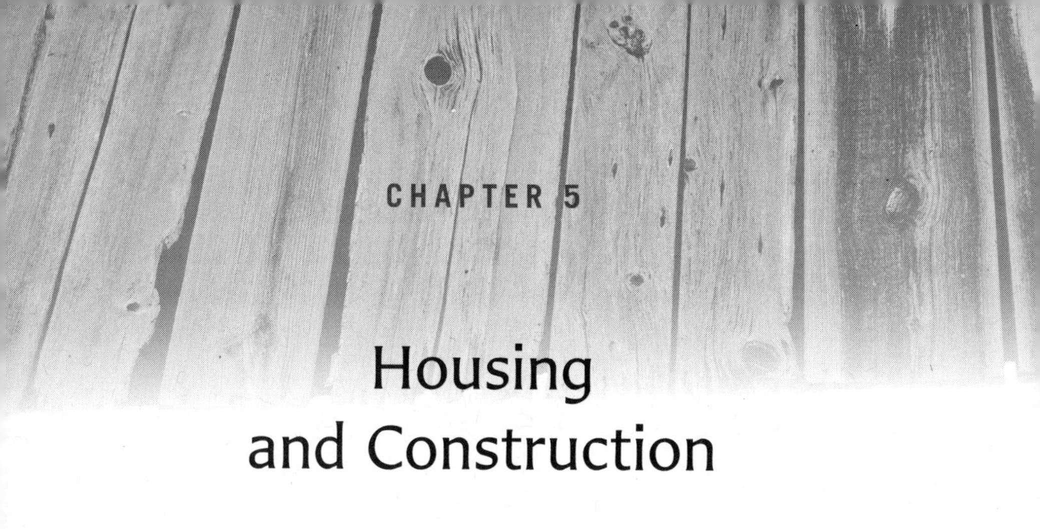

CHAPTER 5

Housing and Construction

SHELTER IS A NECESSITY OF LIFE and one of the core elements for the foundation of most communities. Shelter and housing affect how we live and how we live together. Unfortunately, there is no one-size-fits-all answer to define what type of housing is best. We do know that an economy based upon overconsumption and a continuous housing boom is unsustainable, as proven by the market crash in the US of 2008. Despite a push for greener lifestyles, the US construction industry continues to be enamored by oversized McMansion energy hogs, the source of the greatest profits. Something has to change.

Developing affordable, comfortable housing remains one of the greatest challenges for The Farm, and this relates directly to the community's ability to grow and expand. Decisions made from the initial arrival on the land still shape the community today and influence its future development. Like every aspect of The Farm, housing is infused with history, innovation and pragmatism, interwoven with the unique complexity that is an integral part of life in community.

Moving to the Land

The bus trip across the country prepared the counterculture pioneers from San Francisco for the next phase, the founding of The Farm Community. Living on the road for a year, these new-age gypsies gradually grew used to living without electricity, running water and indoor amenities such as a shower and toilet. After their conversion to bus life before arriving in Tennessee, settling on the land was like finding the ultimate place to park.

Old logging roads, little more than wide dirt trails, dissected the center of every ridge, sprawling in all directions like fingers from a hand. One at a time, the buses would navigate through the ruts until they came to the end of the line, with the first bus to travel down each road establishing the farthest outpost. Then the next bus would follow, seeking a place to pull in, adhering to the general agreement that they should be unable to see each other, allowing each inhabitant to feel that they were in the center of a remote wilderness. The lush vegetation of Tennessee made this easily achievable. Conforming to the natural lay of the land, the buses were usually 150 to 300 feet apart, with some neighborhoods more closely settled than others.

This meant that, in a matter of days, several hundred people were in place, living on the land. There wasn't a lot of forethought about what would constitute an ideal house site or how the community would provide services to these remote locations. The flower children had found paradise and were living the dream, back to the land!

The buildings on the new land consisted of one small ranch-style single-story home and a couple of barns. "The House" had the only running water from a nearby well, which served the home's kitchen and bathroom. It had electricity and a phone. The House gave the group a place to start, but the rest of the community would have to be built from the ground up.

Freedom of Expression

One of the primary benefits The Farm gained when it settled in Tennessee was the total absence of building codes. Back in California, many a commune was wiped off the map by county officials and bulldozers using building violations to eliminate hippie hovels from the hillsides. When the community's founders arrived, the rural poor of

Tennessee lived in some pretty ramshackle structures. The Farm was able to establish itself in buses and tents without the local government even blinking an eye.

When on the road with the Caravan, each bus would transport and supply sleeping space for many people. Now that the buses were acting as permanent dwellings, they were only suitable to house one couple, and even then, the quarters could be a bit confining. The population needed room to move.

People began using whatever materials they could get their hands on to scrounge together simple shacks and lean-tos. Often a bus would have an addition built onto the front entrance, rudimentary structures for storing firewood and tools. In their more developed form, these add-on sheds became outdoor kitchens and living areas, so that the bus itself only needed to be used as a bedroom.

Recycling and Salvage

The demand for building materials was overwhelming, and money was in short supply. What The Farm did have in abundance was labor, strong and eager young people ready to take on the most challenging projects. The community soon realized that one of the best ways to acquire seemingly no-cost building materials was through salvage. It was a win-win situation. Local Tennesseans would have a barn, an old house or a building they wanted torn down. They would contact The Farm, offering the community the chance to tear down these structures in exchange for the materials. Work crews were dispatched armed with crowbars, sledgehammers and gloves.

Tents—"Hents"

Desperate for a quick solution, community members soon discovered that army surplus tents could be purchased cheap and were readily available by the truckload. Measuring 16 feet by 32 feet, these dark green canvas enclosures satisfied the need for both instant housing and community space. With the Wrecking Crew operating in full swing, loads of lumber were being delivered daily, and used to assemble frames that would give the tents a more solid means of support.

Little by little, the tents were converted into "hents," part tent, part house. An army tent's only openings were slits in the canvas on either end, so the inside was virtually pitch black. The solution was simple: cut holes in the walls and install windows, also in ready supply thanks to the Wrecking Crew. A few more imaginative souls even tried cutting out squares in the roof to create makeshift skylights, but this turned out to be a bad idea. No amount of duct tape or tar could prevent them from leaking.

Doors were also added in place of the slits, providing a more civilized entry, while also helping to keep out critters, wandering raccoons or possums in search of a free meal. Eventually salvaged tin replaced the canvas roofs. Interior framing divided up the space into bedrooms and a living area. It was kind of like staying in the ultimate kid club house. Most of the tents were set up as communal households for several couples and a few single folks. Some were designated specifically as single men's or single women's scenes.

Living in the army tents gave real meaning to the sense that this early phase of The Farm was like boot camp. The dark green canvas soaked up Tennessee's hot summer sun, and the inside of an army tent was like an oven, unbearably hot. Tennessee winters can go down into the teens, occasionally below zero, and the thin canvas offered little protection from the freezing temperatures. A woodstove could go through many cords of firewood in an effort to produce heat that disappeared if you moved only a few feet away. It was not an easy life.

The First Public Buildings

The Farm's general population consisted primarily of college and high school dropouts from middle-class America. Few had any real-world skills or experience, especially when it came to construction and the building trades. Anxious to learn, some found work in Nashville and surrounding towns, essentially becoming apprentices, learning how

Based on the shape of a mandala, the 8 sided The Farm Store was one of the first buildings constructed on the land.

to work with lumber and properly frame a home or building. After a time, these newly acquired skills could be brought back to the community and used to build the town.

The first real buildings to be constructed were those that served the entire community. Using plans drawn up by the US government, a three-level building that stair-stepped down a hillside was constructed to process sorghum molasses, a product that could be sold as a cash crop and supply the community with a locally produced sweetener. The massive Tractor Barn was a central base of operations for the Farming Crew. Another building, the Motor Pool, was also like a huge barn, with a cement floor and large garage doors for servicing the expanding fleet of vehicles.

(It might be helpful to mention here that, in adherence to community's early affinity to Zen Buddhism, the proper names of most buildings, roads, businesses and more were always boiled down to their most generic form. This explains the capitalization of the names of buildings like the Tractor Barn, the Motor Pool, or The Farm itself. It is a practice that still holds true today.)

Building The Store was definitely a labor of love. To celebrate its importance as a centerpiece of the community, it has an octagonal design, modeled after the mandala that graces the covers of Stephen's first two books, *Monday Night Class* and *The Caravan*, making this shape the community's first unofficial logo. The building was unique and beautiful, but simultaneously extremely labor intensive. Hundreds of difficult angle cuts were required for the framing and at every other step of the way. It was far from practical, and at the end of the project, the carpenters announced, "No more round buildings!"

Passive Solar

A passive solar building faces south and collects warmth as sunlight passes through glass or some other material. As members of The Farm began to build permanent homes and public buildings, whenever a site had proper southern exposure, some method of passive solar was

The passive solar design of The Farm School allows the low elevation winter sun to pass through translucent panels and warm its cement floor and walls, creating a huge thermal mass.

often incorporated into the design, such as attaching a greenhouse to the front. The most innovative design was encompassed in the 60-by-200-foot Solar School that became the largest passive solar building in Tennessee.

From the side, the roof of the Solar School can be viewed as a series of four sawtooths, each one extending an additional 12 feet above the top of an 8-foot wall. This broad surface is sheathed with translucent fiberglass panels facing the sun. Light passing through the panels strikes a cinderblock wall at the back of each classroom. The heat absorbed by this wall is transferred down to a cement slab floor, turning the entire building into a giant thermal mass that holds and then slowly gives off heat on into the night. As a backup for winter weeks with no sun, a wood-fired boiler was installed in a basement that sends hot water through a radiator system in each classroom. The entire radiator system was assembled from salvaged materials.

Every sawtooth roof line has an extra-long overhang or soffit. The higher elevation summer sun is blocked by the overhang, shading the translucent panels so that sunlight is unable to pass through and bring unwanted heat into the building. The translucent panels also function as skylights, filling the building with abundant ambient light.

Overall, the design has both positives and negatives. Because the sun rises later on winter mornings, it can take several hours for heat to accumulate and warm the building. The teachers and staff have found it better to begin school at 10 AM rather than 8 AM as is typical in public schools. To compensate, classes run until 4:00. In spite of the extended overhang that obstructs the summer sun, the block-and-brick building is still a large thermal mass, and over the course of the summer, it absorbs heat and becomes quite hot inside, so air conditioning units have been added to the classrooms and offices. At the other extreme, when experiencing weeks without sun, the high vaulted ceiling created by each sawtooth makes the rooms difficult to heat. Since heat rises, it takes a good amount of time for heat to build up under the 12-foot ceilings before making its way down to floor level.

Perhaps the biggest problem has been the sawtooth design itself, where the bottom of each roof comes in contact with the next tooth's vertical wall. During torrential rains, a tremendous amount of water flows down the steep roof slamming into the valley where the roof and vertical wall intersect.

The impact of the water caused it to run up the face and over the top of the flashing, which was meant to keep water out. Over the years, every valley leaked. Patch jobs and surface repairs could not stop the water from getting through. Finally, in 2005, each valley was reworked with much more extensive flashing, which helped but did not completely solve the problem. Untold amounts of man-hours and money have been spent in an effort to improve upon the original design.

The sawtooth design experienced a bit more success on a building just outside The Farm. In the early '80s, the mother of a community resident hired The Farm Building Company to construct a solar home. This time, there was only one solar face, so there was no valley from a second sawtooth.

A greenhouse stretches across the entire south-facing wall. Behind the greenhouse, picture windows allow sunlight to enter the home. Heat is absorbed by the cement slab floor that extends under the entire structure. The north wall is protected by an earthen berm, insulating the home from weather extremes, while forming a mass that helps maintain an even temperature. A natural cooling system was installed in which tubes were buried into the earth to bring cool air into the home. Unfortunately, this system was no match for the intense heat of Tennessee summers, and a geothermal cooling system was later installed.

Brick and Steel

The early work of the salvage operation has had a major impact on The Farm's architectural landscape. One of its biggest jobs was the demolition of a three-story shirt factory, with brick walls three layers deep, just off the town square in Pulaski, Tennessee. Countless bricks

The large dome adjacent to The Farm Store is constructed from steel salvaged by community work crews in the 1970s.

were brought back to The Farm and used on the exterior of several prominent public buildings.

In the nearby town of Columbia, a church wanted space for a parking lot, and the building in the way was a round-roof warehouse. Supporting the roof was a series of steel trusses 60 feet wide. A crane was brought in to lower the trusses onto a trailer pulled by one of The Farm's semi-trucks. In the early '80s, these trusses were arranged in a circle to form a large dome. The concept behind this Meeting Hall was to build a space large enough to enclose the entire community population of over one thousand. By this time, The Farm had several skilled industrial welders in its midst and had its own crane. Large steel arches were assembled from more salvaged steel and rebar, joining the trusses together and completing the dome.

The plan was to cover the structure with mesh and concrete, a treatment known as ferrocement. Before this could be accomplished, The Farm's financial crisis of the 1980s struck, and construction stopped. For over two decades, the framework stood in place, functioning as an oversized gazebo, and became a community icon. National Guard helicopters and airplanes flying overhead used the dome as a navigational landmark.

In 2009, the structure was covered with an industrial-grade canvas, and ideally at some point, rubberized coating will be added to extend the cover's lifespan to an estimated 40 years. This space at the center of The Farm's downtown shelters a playground. It has become a much-appreciated gathering place for music, picnics and other events.

After The Changeover

By the early '80s, in addition to the multitude of tents, shacks and buses were still being used as housing, The Farm had over a dozen medium and large houses built entirely from recycled materials and salvaged lumber. Each home was a commune in its own right, providing shelter for 30 to 45 people. The places were seriously overcrowded, with small bedrooms for couples and all of the children bunking together in communal kid rooms.

After The Changeover, people still living in tents and other types of inadequate shelter were more likely to leave, while those in one of the real homes that had been built during the communal period stayed. For the first few years, there was some shifting around as people continued to depart until each house had only one family, with a few of the homes converted into duplexes.

All of the houses needed a lot of work. Windows were old and leaky or even nonexistent, with plastic stapled over where glass should be. The roofs covered by recycled tin from old barns frequently leaked. Interior walls were unfinished, with bare sheetrock that needed plaster work and paint. Only a few houses had exterior siding. Once the communal economy became a thing of the past, it was up to the people occupying each home to earn their own money to make any improvements.

One aspect that did not change was the ownership of the land.

It remains in a trust to preserve the integrity of the community, and it is universally regarded as the glue that holds The Farm together, preserving its identity and cohesiveness. This means that people do not legally own or have a deed to their homes. If the land became divided into dozens of separate small deeds, the community would not have the ability to control membership. Homes could be sold and purchased by anyone, and over time, the population could consist of people with no relationship to the principles for which the community was founded or the agreements it maintains today.

Because The Farm's members do not own their homes, it can be difficult to get a home improvement loan from a bank. For the most part, those who wish to make upgrades and additions will save money and then invest in the next priority, whether it is a new roof, adding a bathroom or whatever they feel would make the most difference in their lives. As families and individuals began investing their own money into home improvements, there was general agreement that this investment represented equity. This equity becomes the value or cost of a home if it changes hands, or is "sold" by one member to another.

Anyone wishing to build a new home is unable to get a standard mortgage because the bank cannot foreclose on a defaulted loan and put the house up for sale on the open market. There is unanimous agreement that the community's assets are not to be used as collateral with a bank in order to provide financing for mortgages and new home construction. After paying off the massive debt incurred during the communal period, Farm members do not want to burden the community with debt or do anything that could give banks leverage or in any way endanger the land.

Over the decades since the Changeover, the community has been able to generate its own capital, allowing it to provide members with small home improvement loans, but it has not accumulated the hundreds of thousands of dollars in liquid capital necessary to finance the construction of multiple new homes. The purchase of equity or financing new home construction remains one of the community's biggest challenges, limiting growth and making it a real hurdle for potential new members seeking to move to The Farm.

But where there is a will, there's a way. Since The Changeover, numerous new homes have been built, and usually several home construction projects are underway every year. These homes illustrate the ingenuity and determination that are required when people join the community. The solutions have been many and are as varied as the people themselves. No one claims that the community has figured out a fair and just system or that it is an easy process, only that people have found ways to make it work within the constraints of the system in place.

Many of the new homes have been built by former members who lived on The Farm during the communal period and left at the time of The Changeover. They understand the importance of maintaining the integrity of the land and are in tune with the community's values. They are willing to forego legal ownership of their home and make an investment that does not deliver the same economic security as the purchase or construction of a home on the outside. They place a great value on the shared values and benefits of life in community and are willing to do what it takes to build or acquire a home.

A few people have moved back to The Farm and built a home after they received an inheritance. For others, life outside The Farm has given them access to higher-paying jobs. They purchased homes that increased in value. Before coming back to The Farm, they were able to cash out, converting that home equity into money used to build a new home on the land. At times this means starting small, building a cabin just large enough to supply the minimum amount of space needed to be comfortable and then expanding a few years later as finances and savings permit.

The cost of building a home is always a concern, and one of the best solutions continues to be the owner-built home. Like everywhere else, the cost of labor is much greater than the cost of materials, and those who are able to do even a portion of the construction themselves have a distinct advantage.

Another way people have found to keep new housing affordable is to utilize kits, prebuilt cabins and other types of manufactured dwellings. Buildings manufactured off-site are able to reduce costs through more efficient means of construction, like assembly-line mass production, bulk purchase of materials and consistent designs that can reduce both material and labor costs, passing those savings on to the buyer.

Middle Tennessee has a sizable Amish community as well as New Order Amish, a group with similar religious beliefs that allows their followers to have access to certain aspects of modern technology like power tools. Both groups have a lifestyle that allows them to live well below the typical American's standard cost of living, enabling them to be competitive and cost-effective builders and contractors. A market has developed for inexpensive cabins built with Amish labor.

The kits and cabins are sold as unfinished shells consisting of walls, a floor and a roof. They come complete with windows, doors and exterior siding. It is up to the buyer to finish the inside, adding electrical wiring and plumbing, insulation, and interior walls, plus all of the components needed for a kitchen and bathroom. A number of kits and Amish cabins have been set up in the community to serve as homes and guest cabins, with finished construction costs coming in well under the usual cost per square foot of standard homes. It can be an inexpensive way to get established on the land.

Manufactured homes and prefab commercial buildings, used on The Farm in several locations, have one very distinct advantage: bank financing. Because the structure is designed to be moved, if necessary the bank can repossess and remove the building, recouping their investment. These buildings are ready to occupy when they arrive onsite, complete with kitchens, bathrooms, even heating and cooling systems. Great improvements have been made in recent decades in the quality of materials and construction, but in general, manufactured buildings do not have the level of durability or energy efficiency of a custom-built home. From a financial standpoint, they provide great value in terms of the cost per square foot and the principal investment. On the other hand, this type of building will typically have higher heating and cooling costs and require additional maintenance where corners were cut by the manufacturer to keep the initial cost down.

The Deltec Home

The Deltec is a 16-sided round house kit, promoted as "The Original Green Home." Its components, including roof trusses, floor system and prewired wall panels with exterior siding, are all manufactured with green energy generated by solar panels at the company's facility in North Carolina, and then delivered and assembled onsite.

According to the company's literature, round homes are more energy efficient, exposing less outside surface area while maximizing the amount of enclosed square footage. Improved aerodynamics and tight seals means there are fewer drafts, with improved overall comfort along with lower heating and cooling costs.

This round Deltec home was assembled from a kit delivered from the Deltec manufacturing facility in North Carolina.

The Deltec home on The Farm incorporates passive solar design with several windows installed on the south side of the home. A two-foot roof overhang helps block out the higher-elevation summer sun, while allowing the lower-elevation winter sun to pass through the windows bringing warmth and sunlight into the interior. In addition, light tubes were installed to bring passive solar lighting into the interior rooms. All in all, it is a beautiful and unique living space.

Green Building and Natural Building

The definition of "green building" can be very broad, and it can be eye-opening to learn what constitutes "green" in modern construction. All people who want to live in a more sustainable way find they must make compromises, and this is no different on The Farm. It is up to each member to make responsible and, at the same time, practical choices, decisions affected by both their pocketbook and their situation. At the heart of this is the contrast between green construction and natural building.

Green building generally includes anything related to energy efficiency and renewable or sustainable building materials. A home or structure may be built primarily from conventional materials and construction methods but incorporate numerous features to improve energy efficiency.

Natural buildings are made from materials that are locally sourced and require little to no manufacturing or processing. Earth, clay, sand and straw are combined with locally harvested timbers in a variety of ways to create a structure. Materials costs can be low, but significantly more labor is involved. This can raise costs to be as much or more than conventional construction, often making natural building projects best suited for homes built by the owner, someone willing to commit their own time as a labor of love. The majority of The Farm's homes and buildings would fall into the green category, but natural buildings are here and there, the largest concentration on the campus of the Ecovillage Training Center. As more people have the opportunity to see how the costs, benefits and maintenance of natural versus green buildings compare over time, others may be encouraged to go the natural route as well.

Another consideration relating to both green construction and natural home building is toxicity. Commonly used plywood and particleboards are held together with chemically potent glues and adhesives. Paints, polyurethanes and sealers may contain all sorts of chemicals outgassed into a home's interior air. Green contractors make an effort to choose people-friendly alternatives, while natural builders avoid this altogether by working with unadulterated materials that come directly from the earth. The question of toxicity goes beyond a material's effect once it reaches the job site to include its impact on the planet from the acquisition of raw materials through every stage of the manufacturing and delivery process.

The building of a home can represent a sizable investment. Labor, the cost of hiring a contractor and building crews, represents the largest expense, far outweighing the cost of materials even with conventional construction methods. Builders are most efficient when doing what they know, relying on their experience, which frequently dates back to many years of work with standard materials and established home designs.

Although humankind has been building homes with natural materials for hundreds, even thousands, of years, today this type of vocation falls outside the mainstream. Most people working in construction are unfamiliar with natural building materials and techniques. Those who do have this experience are specialists, which can mean they will cost more than a standard construction worker.

Even relatively straightforward and easily understood building methods such as post and beam construction still require more time and labor than it takes to build a same size structure using standard commercially available framing lumber. Other aspects of natural building include sourcing materials, locating trees, cutting them down and removing them from the forest, transporting logs to a sawmill and lumber back, curing to ensure it is not green and has gone through any shrinking; all of these come at the cost of time and labor.

It is difficult to compete with the efficiency of placing a single phone call to a building supply company and having all of the project's materials delivered by truck directly to the job site.

Being ahead of the curve is never easy. The people willing to put in the time, effort and extra financial investment to craft examples of natural building are appreciated as leaders helping to fulfill The Farm's mission, to be a model of sustainability.

People building or having a home built for them must weigh all the pros and cons. They must balance a desire to express their ideals regarding sustainability and how it applies to home construction in this day and age, while working within the confines of a limited budget. For most of us, no matter what style of building we choose, it must also be affordable.

In this way, The Farm is like a living laboratory. It is possible to step back and observe which choices members of The Farm have made and why, examining various aspects of home construction, looking at the conventional approach, green options and the natural building alternative.

Energy Efficiency and Insulation

A common concern is the importance of energy efficiency, generally regarded as a reduction in power consumption, particularly as it relates to heating and cooling costs. One of the primary ways to accomplish this is through insulation. The type of insulation used in Farm homes has a direct correlation to the timeline and progression of technological advances and their availability in the marketplace.

Insulation is rated by its R-Value per inch of material, representing the resistance to heat flow, keeping heat out in summer and holding onto it in winter. A 6-inch wall would have a 50 percent higher R rating than a 4-inch wall of the same substance. Following that logic, it would take twice as much of material rated at R-3 to achieve the same level of insulation as it would for a material rated at R-6. These numbers are important when designing a home in order to estimate its energy efficiency.

For many decades, the only commercially available insulation was fiberglass, having a value of R-3.6 to R-5. Fiberglass is manufactured from sand and 20 to 30 percent recycled glass that is combined and spun into a light and fluffy material resembling cotton candy. The sand is normally acquired through mining, and the fiberglass manufacturing process includes toxic chemicals such as formaldehyde. The spun fiberglass is often attached to a paper or foil batt that can be stapled to wall studs for easy installation, simultaneously creating an air and vapor seal. Holes cut to provide access to electrical boxes and light fixtures break that seal and can be a source of cold air flow or drafts. Fiberglass is also the least expensive option, and it is still widely used in standard construction.

Insulation made from cotton scraps, such as the leftover material from the production of blue jeans, was one of the first readily available type of green insulation. It consists of 80 percent recycled material and does not contain formaldehyde. Acquisition of the raw material (farming and recycling) has less impact on the environment than mining. The R-value is roughly equivalent to fiberglass, but installation and overall costs are frequently a bit higher. It is sold in rolls and stapled into walls like fiberglass or blown 6 to 8 inches deep into attics and ceilings.

Blown cellulose made from 80 percent recycled newspapers is another commonly available green insulation that carries a value of R-3 to R-4. Cellulose initially lost favor because it had been found to settle as much as 20 percent over time, reducing its insulation qualities. This was addressed by increasing the volume or density of the material, making blown cellulose a viable alternative, especially in remodeling, using it as a filler in existing walls. Both cotton and cellulose insulations are treated with chemicals as a fire retardant.

Spray foam is a liquid that expands on contact, filling all crevices or gaps producing a perfect, tight seal, blocking any intrusion by outside air. Unlike blown cellulose, it adheres to any surface, allowing it to be used between rafters, floor and ceiling joists or against lightweight skirting enclosing a pier foundation. It has a value of R-3 to

R-7, as good or better than any other insulation. The foam also adds to structural stability, functions as a sound barrier and works better than plastic as a vapor barrier. The downside? Spray foams have been petrochemical based, but there are formulas based on soy oils and others are being developed from recycled sources. The HFC (hydrofluorocarbon) and HCFC (hydrochlorofluorocarbon) compounds used as propellants to spray the foam are considered to be greenhouse gasses. Foam also releases toxic fumes if burned, such as in a house fire. Hopefully both of these drawbacks will also change as newer methods and formulas are developed. Because of its excellent R values and ability to seal any and all intrusion from hot or cold air, spray foam insulation has been the choice in many newly constructed homes on The Farm.

Structural insulated sheathing is a foam panel measuring 4 feet by 8 feet and 1 inch thick. These are often installed under siding on an exterior wall or tin on a roof as an easy way to gain up to an additional R-6 of insulation. One side of the foam core panel is coated with foil, reflecting heat while also effective as a moisture barrier.

Insulated Concrete Forms (ICFs)

Several new homes have been built using Insulated Concrete Forms or ICFs, another variation on the use of foam for insulation, in this case, combining insulation with the structural support for the building. ICFs consist of two thick polystyrene foam panels held together by plastic ties. The panels are stacked like big building blocks and assembled into walls. The hollow forms are then filled with concrete. ICF buildings are incredibly solid and virtually tornado-proof, providing added security since The Farm is only a few miles away from a tornado alley.

If the home has two stories, the bottom section uses an 8-inch-wide ICF and the upper floor a 6-inch form, permitting it to rest perfectly on the wall below. ICF homes are extremely insulated, requiring minimal energy to heat or cool. Insulation R-values go up as high as 26 when factoring in both foam panels and the cement. The panels

(Before) Homes built from insulated concrete forms (ICFs) are assembled from foam panels sandwiching 6 to 8 inches of cement, making them tornado and earthquake proof, and extremely insulated.

(After) The finished ICF home.

are quick to assemble, saving on labor, but overall ICF homes normally cost approximately 8 to 10 percent more than a conventional wood-frame structure. The interior and exterior foam panel walls can be covered with a variety of building materials.

Straw, the Natural Choice

The principal insulation material used by natural builders is straw, a locally available, renewable resource that can be utilized in a variety of ways. It eliminates the manufacturing costs and extensive technology necessary for other common insulation materials such as fiberglass and the increasingly common spray foam.

Straw bale construction has become a popular form of green building around the country, more so in the arid climates of the West, but popping up east of the Mississippi with increased frequency. One of the oldest examples of straw bale construction in the South was built in 1938, about an hour away from The Farm in Huntsville, Alabama.

The Farm has one (very nice) straw bale home. The main structure and roof are supported by locally harvested cedar posts and

Straw is a readily available, inexpensive and renewable material that can be utilized in a number of ways to insulate a home.

beams. In between the posts, bales are stacked like building blocks, producing walls approximately 24 inches thick. This gives them a total estimated R-value of R-25 to R-40, keeping the home warm in winter and cool in summer, with minimal dependence on heating or cooling systems.

It is critical that the roof integrity of a straw bale home is perfect and continuously maintained, eliminating any possibility of water coming in contact with the straw. When moisture is allowed to penetrate, mold and mildew can follow soon after. A straw bale home built near The Farm had to be demolished because a poor foundation had settled and allowed moisture to intrude from the bottom. As the mold and mildew took over, the home simply became unlivable.

Another use of straw as a building material combines it with liquefied earth, or clay slip, and there are several examples of its use in the community. R-values for straw clay slip are modest, so walls are usually built 6 to 10 inches thick to compensate.

Compacted bales of straw are broken open and fluffed before immersing the loose straw in a tank of water and clay. The mixture is stirred in the slurry until each blade is completely coated. One possibility is to use a repurposed cement mixer to perform this task.

Post and beam or conventional wood-framed walls are encased with forms, using plywood or particleboard to create deep channels. The wet coated straw is stuffed or tamped into the forms with enough pressure to ensure that the wall is a solid continuous mass. The forms are removed as soon as possible so drying can begin immediately. If need be, the forms are then raised higher, reattached, and the process is repeated until the walls are complete.

The straw clay slip is then given time to cure or air dry until every bit of moisture has evaporated. Depending on the air temperature and humidity, the curing period can last several weeks to a few months. During this time, any kernels of wheat or grain will commonly sprout with tufts of green emerging from the walls. Deprived of any additional moisture, the sprouts will eventually dry up and do not present a problem.

More Insulation Options

Of course, there are other factors to consider when designing a home for maximum energy efficiency. In the past, windows have been a primary source of heat loss in winter and gain in summer. Today's low-e (emissivity) windows have a metal oxide coating that reflects heat but lets light pass through. The space between dual-pane windows is filled with argon gas, superior to air alone as an extra layer of insulation. These features will increase a window's price but have become one of the most important methods for any home to improve its energy efficiency.

In Tennessee, the concern is as much, or more, about keeping the home cool in the summer as it is about gathering heat in the winter through passive solar. Rather than design for maximum solar gain in winter, a different tactic can be to reduce the number or size of windows on the south wall, letting in some light while minimizing heat absorption. When appropriate, windows or a sliding glass door installed on the north wall can be opened during the summer to let in cool air.

Adobe, Cob and Earth Bag

Most homes in the US are constructed from pine lumber for a standard method of building nicknamed "stick framing" after the dozens 2-by-4 boards that make up the main support. Pine trees are a renewable resource, but massive pine plantations often displace native forest species, and many experts consider them to be an ecological disaster. Monoculture plantings leave pine plantations susceptible to insect infestations and other biological disasters.

Demand for timber continues to drive the destruction of the world's last remaining virgin forests, which have disappeared in the US and are vanishing at an alarming rate in their last remaining stands around the world. The 2-by-4s and other lumber delivered so innocuously to your building site have likely traveled thousands of miles. The cost of lumber skyrockets while quality plummets. The entire system from beginning to end is clearly not sustainable.

In contrast, a key principle of natural building is that its materials are locally sourced. For thousands of years, humans have built their homes from the materials around them. As oil prices rise and transcontinental transportation becomes increasingly cost prohibitive, we may well see a return to the geographical relevance of sourcing all materials locally.

Adobe is one of the oldest building materials known to man, another variation on clay, sand, water and some amount of straw or manure. Because there is no reinforcement tying dried adobe bricks together, like rebar in cement, the resulting walls don't have the same stability, especially problematic in earthquake-prone regions like Central America where adobe is still widely used.

Very similar to adobe, cob is a building material that is also centuries old but seeing a new popularity in green building circles. It is easy to work with, requiring few tools. Natural builders are drawn to its low cost and the flexible, imaginatively shaped buildings that can be fashioned. With a higher content of straw, cob is said to be an improvement over adobe. The long fibrous blades add greater strength, holding the material together. The mixture, with its consistency like bread dough, is similarly formed into loaves, stacked in rows like bricks while wet to form the walls. Sticks are periodically shoved down vertically through several loaves, tying the separate rows together, much in the same way rebar is used in cement, adding to the stability.

Earthbag building was inspired by sandbags used for floodwater control and military bunkers. In its current usage, polypropylene bags manufactured as grain or feed sacks are filled with an earth and clay or adobe mixture. Here the bags become the bricks, again stacked in rows, layer after layer. Often a strand of barbed wire is laid down between the rows as a way to bind them together. The sacks can also be pierced with rebar to join rows together, especially at corners and junctions. When purchased on long rolls, the bags can be cut to any length, encouraging more variation in shape and design. With careful construction, concentric earthbags can even be gradually sloped inward to form domed roofs.

Working with Wood

One excellent example illustrating the use of both local and recycled materials would be a log cabin on The Farm that had its initial construction before The Changeover, with significant changes and improvements taking place after that. Cedar logs comprising about half of the structure came from an 1800s Nashville area home salvaged by Farm work crews. The other logs came from oak trees harvested to open up house sites on The Farm. The logs were hand hewn into shape by a member of The Farm who became one of the cabin's first residents. Many of its support beams are also from timbers harvested from the building site.

Cordwood walls are constructed of locally sourced logs in a variety of sizes, filling the spaces in between with earth based plasters.

The log walls are about eight inches thick. Because wood is comprised of plant cell walls, the millions of tiny air pockets make an excellent source of insulation, keeping the cabin warm in the winter and cool in the summer. Over the last 30 or so years, the cabin's log exterior has received occasional preservative treatments and stains. Some of the interior walls have been sanded and linseed oil applied to bring out the natural honey color of the oak. With The Farm's extensive forests, log cabins might seem a logical and cost-effective way to acquire more housing, and someday this may be the case. For now, log homes are more expensive to build, and this approach has not been undertaken by any members since The Changeover.

Cordwood construction is a much different way of working with logs and yet one more method of natural building. The term "cordwood" refers to the length of the logs, approximately 18 inches, and the way they are stacked to form a wall, essentially much the same as a cord of firewood. Logs from poplar trees, a native hardwood and an abundantly available species in Tennessee, are ideal because they are extremely light, which means the interior cells of the wood are filled with air, making them an excellent insulator, especially when the walls are 18 inches thick.

Small and medium-sized trees harvested locally are the basis of round pole construction, another slant on natural building. The logs and poles can be used as posts and load-bearing structural supports, or as floor joists and rafters. A draw knife is used to peel away the bark. The wood is left to age or cure so that it does not shrink after it is put in place. The irregular shape and size is a departure from the straight lines and hard corners that mirror the rigid conformity of modern building materials.

Exterior Siding and Sheathing

After The Changeover, virtually every home on The Farm needed exterior siding. New home builders are also faced with the same decision: Which exterior siding represents the best choice in terms of affordability, durability and aesthetic appeal?

Historically, homes in the South were covered with wood. Familiar lap siding runs horizontally across a wall, with each piece overlapping the one below. Because most wood siding required paint every few years to protect it from weather and rot, in the '70s, an aluminum version was developed and became quite popular, but was too expensive to be an option for use on The Farm. The cheaper alternative was siding made from a fiber composite, which was installed on several public buildings. When aluminum became cost prohibitive, it was replaced in the marketplace by vinyl siding. Relatively inexpensive, vinyl has become one of the more prominent types of exterior siding across the country. Essentially a form of plastic, vinyl never needs painting, but can become brittle after years exposed to sunlight and would in no way be considered "green."

Cypress and western cedar are both woods milled for exterior lap siding that are naturally resistant to rot and insect damage. After installation, an exterior sealant can be applied to help protect them from moisture. These woods are often chosen for houses on The Farm due to their warm amber color and rustic appearance. The color does fade and turn to grey over time and can even turn black from mildew that grows on the wood in Tennessee's humid climate. After about 20 years, the cypress siding of one home was pressure cleaned with a bleach solution and given another treatment of exterior sealant, restoring the wood to its original color. Milled and sold by the local Amish, the cedar and cypress logs are actually brought from states farther west and in most cases are not locally harvested in Tennessee.

Board and batten siding consists of vertical boards 8 to 12 inches wide nailed to a wall, with the space between each of the boards covered by a 2-to-3-inch narrow strip of wood, the batten. A number of new homes have chosen this type of siding, using locally harvested poplar boards. They also present a rustic natural appearance that matches the character of back-to-the-land living.

Exterior siding or paneling sold in 4-by-8 sheets known as T 1-11 is usually designed to resemble vertical board siding and can be found on homes and buildings throughout The Farm. Manufactured as plywood or from composite wood fibers, the large sheets are relatively inexpensive and go up quickly, preferred when keeping costs low is a priority.

Cement or Hardie board is a composite of cement and wood-based or plant cellulose fibers, sold in both 4-by-8 panels or as lap siding one-half-inch thick. It is resistant to fire and moisture, will not mold or mildew and is virtually indestructible. The internal fibers make the material very porous, readily absorbing paint that does not peel or need repeated applications and continual maintenance. Due to its weight and hardness, the siding is commonly installed by pre-drilling holes and hung with screws. This makes it a little slower to install than other types of siding, but the extra labor expense is offset by the maintenance-free, long life span. All of these qualities combined make cement board a desirable choice by contractors and builders going green, and it has been installed on numerous homes and buildings across The Farm. On the other hand, the extensive processing necessary to manufacture cement pushes it outside the criteria for materials acceptable for use in natural building.

Nevertheless, cement was used for both the inside and outside walls of The Farm's straw bale home, applied like stucco. A cement truck outfitted with a special pump sprayed the wet cement onto the surface of the straw bales. The rough texture of the bales allowed the cement to easily adhere, embedding nearly an inch thick into their surface. About 40 members of The Farm working as volunteers used trowels to smooth the surface inside and out, with a mason from the community adding some bas relief artwork as a finishing touch. All of the cement work inside and out was completed in a single day. As the cement began to dry and cure, cracks began to form in places such at the corners of windows. To prevent more cracking, expansion joints were added by cutting grooves in the cement with an electric saw and concrete blade.

Earthen plasters and stuccos are the natural alternative and the logical choice for covering the exterior and interior walls of straw bale, straw clay slip and other types of natural buildings. The colors and tones inherent in the earth itself let the buildings blend seamlessly with their surroundings, while eliminating the need for paint or other types of chemical-based wall coatings. Unlike other materials typically used for interior and exterior walls, earth plasters remain porous, permitting the building to breathe. In addition, proponents of natural building state that moisture and humidity from within the home will transfer from inside to outside, acting as a natural and organic cooling system.

The preparation for an earth-based plaster begins by running raw earth through a series of screens, sifting out any rocks, pebbles or unwelcome objects, resulting in a fine powder. The sifted dirt, along with sand, lime or other ingredients, is mixed with water to achieve the consistency of stiff mud. Even though the cost of materials can be next to nothing, the mud plaster is spread by hand, which is very labor-intensive. A rough layer is applied and then left to cure, followed some days later by a second rough coat. A layer of extremely fine material is applied as a finish coat, producing a smooth and attractive wall surface.

It is essential that homes with earth plaster exteriors are constructed with extended roof overhangs and gutters to prevent any splash back from rain coming in contact with the exterior walls. This protects and preserves the integrity of the plaster and also prevents any moisture from being wicked inside the wall. The good news is that any damage due to weathering is easily patched and repaired with a new application of plaster.

Earth plasters are somewhat experimental. Their texture and other properties will vary and are very much dependent on the point of origin of the earth used. Poor-quality mixtures will be susceptible to excessive cracking. Interior walls may shed a coating of dust on furniture, counters and throughout the home, prompting some to choose earth plasters for the outside and standard sheetrock for the inside. Proponents of earth plasters advise starting with small test batches until a good formula or recipe is determined. Consulting with someone that has experience can be helpful.

Roofs

On the early Farm, most building had roofs of either rusty recycled tin or asphalt roll roofing and shingles. Following The Changeover, shingles continued to be the least expensive option for many years. Tin was mostly reserved for barns and sheds and needed frequent painting. That changed with the introduction of enamel-coated metal roofing. This baked-on coating comes with a 40-year guarantee, much better than the 15-to-20-year life span of shingle roofs, which crack and disintegrate after years of exposure to sun and weather.

Enamel roofs are available in a variety of colors. A red roof with white exterior siding is a popular combination in Tennessee. Those choosing earth tones might go with a charcoal grey or chocolate brown. Green has been used on The Farm to help the home be less obtrusive and blend into the forest.

As Farm builders became more conscious about green building design, they made the connection that a lighter color will reflect rather than absorb heat, which in turn can reduce a home's cooling costs. White enamel roofs work best, with pigments that block UV, extending the life of the enamel. Cool even on hot sunny days, white

Living roofs insulate and naturally cool the home below through the evaporation of moisture.

enamel metal has become the roof of choice for most new homes on The Farm.

Once common around the world, living roofs are experiencing a comeback. They are a direct answer to global warming and can be seen as the crown jewel of natural building. They represent the polar opposite of oil-based asphalt shingles or the mining, processing and manufacturing inherent in metal roofs. During construction, after the roof support system (rafters and sheathing) have been put in place, a rubber membrane, such as the type used as a pond liner, is rolled out as a moisture barrier covering the entire surface of the roof. This is followed by a layer of recycled carpeting and finally several inches of dirt. Roots from the plants growing into the earth reach down into the carpeting, holding everything in place.

Living roofs function as both insulation and a thermal mass. The greenery and the earth absorb and hold moisture, reducing runoff. Evaporation creates a natural cooling system. It is important for the natural builder to ensure that the roof has been properly engineered and is able to support the added weight. During the hot, dry summers of Tennessee, it may be necessary to water the roof to keep plants alive, which can help keep the home cool during the season when it needs it most. One home on The Farm with a living roof has set up a water catchment system consisting of a cistern collecting runoff from the roof during heavy rains. This becomes an immediately available source of water that does not pull from the community's drinking water supply.

Flooring

Flooring is yet one more item that comes with a wide range of green possibilities. Floors take the brunt of everyday impact on a home, making durability of critical importance. They are highly visible and greatly affect the appearance and overall feel of interior living space.

Finished floors can also be expensive. People building new homes on The Farm facing budget constraints have frequently opted to start out with something temporary, planning to add the permanent floor in Phase 2, after they have saved more money. High-quality plywood can be painted and will work as long as necessary, repainted every few years if need be. Some have chosen to go with wide pine boards for their subfloor, coated with polyurethane to give it a shiny finished look. Unlike plywood or particleboard subfloors, the pine does not have any formaldehyde glues, but it is too soft to serve as a permanent floor, susceptible to dents.

For generations, the standard for interior floors has been hardwood. Several homes and buildings constructed during the early Farm days have hardwood floors recycled from gymnasiums. A number of local Amish sawmills specialize in hardwood flooring. Their flooring requires extensive sanding and finishing, something that can be endured with new construction, but is impractical when remodeling, where commercially available prefinished hardwood is the better choice.

Bamboo flooring resembles hardwood and has become popular because it is affordable and, as a renewable resource, is a green alternative. Although marketing materials claim it is just as durable as hardwood, users on The Farm have found that it is actually quite delicate and susceptible to scuffs, dents and scratches. It may be best to use bamboo only in low traffic areas like bedrooms or living situations where there is minimal impact.

Ceramic tile floors are durable, very attractive, easy to install, and on The Farm as elsewhere, you'll find them used mostly in foyers, entryways, kitchens, bathrooms and basements. Several homes went one step further by opting to install tile floors in main living areas; only instead of ceramic tile, the choice was a natural stone, Vermont slate. The slate is delivered by truck on a pallet, precut either into squares or rectangles of various sizes that form a pattern. Its green color brings a pleasing ambiance to any room.

Outdoor decks present yet another opportunity to go green. To combat wood's inevitable deterioration from exposure to the elements, technology came up with chemically treated wood impervious to bugs and rot. Unfortunately, one of the key ingredients in preservatives and pesticides was chromated copper arsenate, otherwise known as arsenic. On the market for over 20 years, it has since been replaced by newer non-toxic wood treatments and expensive deck flooring made from recycled plastic.

The green alternatives are cedar and locally harvested sassafras lumber, also laden with oils and naturally bug resistant. Unlike the sassafras shrub commonly used for tea, the variety milled for decking lumber grows to the size of small trees in the lowland valleys of Middle Tennessee. Several homes in the community have decks, steps and ramps constructed from sassafras and cedar from local Amish sawmills.

Other Examples of Green Building: The Owner-Built Recycled Home

For many years, John collected materials from construction sites and recycle centers, accumulating almost everything needed to build his dream home: floor and ceiling joists, studs for wall framing, rafters, windows, doors, the porch decking. Even the kitchen cabinets were acquired after a Farm contractor did a kitchen remodel for a Nashville home. The downstairs entryway, kitchen floor and a backdrop for the kitchen stove feature beautiful ceramic tile, another material collected for pennies on the dollar. A salvaged large picture window supplies a great view of the outdoors over by the breakfast nook. The living room features beautiful stained glass lighting fixtures that came from a church.

His home is a testimony to what can be accomplished with a little time, effort and a watchful eye, creating something good from the abundant waste so prevalent in our society today.

The Earth Shelter

One of the most innovative and ambitious examples of green construction on The Farm is called an "earth shelter." The back and sides of the structure are submerged into the hillside, using the thermal mass of the earth to insulate and regulate inside temperatures. The walls are built from cinderblocks filled with concrete and rebar (more mass). The entire south face of the building consists of glass. Sunlight entering the building in winter months strikes the cement slab and Vermont slate floor, along with the block walls, warming the thermal mass represented by the entire structure. In addition, a split design in the roof near the peak made it possible to include a row of small windows (also facing south) to help bring in ambient light.

The result: Very little external heating or cooling is necessary to maintain an even temperature inside the building. Even without added heat, the temperature inside changes very little overnight in winter months. On the coldest days, when skies are overcast and there is no sunshine, a small wood stove adds supplemental heat, which again is absorbed by the building's thermal mass.

The Farm's examples of green building represent elements that can be assimilated into mainstream construction in the here and now. If implemented on a national scale, even the smallest changes could make a tremendous difference and impact the environment. It is important to demonstrate that going green is something we all can and need to do in a way more people understand and can relate to. With its many different models, approaches and years of experience, The Farm can be seen as a test site to find out what works, passing on what it has learned to a wider audience looking for ways they too can live green.

Building with natural materials was once a skill set that every person grew up learning, at least those who worked for a living with their hands. As society has become increasingly stratified into specialization, few people carry with them the skills of creating shelter. Proponents of natural building find inspiration in this return to the basics, a physical expression of their values through construction skills that anyone can learn and apply. Using materials sourced locally and directly from the environment surrounding the home carries with it a satisfying fulfillment, a return to the natural order. This can be a definite motivation for young people in this century drawn to life in community.

Rather than incorporating a unifying vernacular, construction on The Farm reflects its position as a living laboratory, responding to constantly changing times, the availability of new materials, and increased knowledge and awareness. At the same time, its members must contend with factors such as affordability, an absence of financial assistance or financing, and the skillsets of its available builders. All of these things exert their influence.

Life on The Farm is an expression of freedom, and one of the community's strong points is that its members are free to choose how they want to live, including their preferred type of housing. Our home becomes a statement and a reflection of who we are, quite literally, our place in the world.

CHAPTER 6

Membership and the Gate

THE FARM WOULD NEVER have survived without the Gate. From the very beginning, the Gate was considered to be the front door to every home in the community. Everyone wants to feel safe in their home. The Gate saved The Farm from being overwhelmed by random people passing through in droves, trading anarchy for organized management and defined membership. Although much has changed, many aspects of the membership process have remained consistent through the decades, as important today as they were 40 years ago.

During the hippie invasion of the '60s, much of San Francisco had functioned as a "crash pad," a phrase truly born of the hippie era. With hundreds of thousands of young people descending on the city, on any given night, apartments all over the Bay area gave shelter to the masses, the runaways, castaways, seekers and every other flavor of flower child. In the midst of this chaos, an understanding of the value of stability eventually rose to the surface.

The people attending Stephen's Monday Night Class on a regular basis had begun to see themselves as a loosely affiliated group, but leaving San Francisco on the buses took this to a new level. People worked and hung out together every day for many months, gaining a sense of identity and community. When the group arrived in Tennessee, Stephen wisely guided The Farm toward establishing ground rules and agreements that defined the relationship of people coming to and then living on The Farm.

It was recognized early on that if the community was to succeed, the people living there had to give it their full commitment. Either you were making The Farm your home and ready to build a life there or you were just passing through. The first point of interface for every person arriving was the Gate.

The Gate was staffed 24 hours a day, 7 days a week, constantly monitoring and documenting every vehicle and person arriving and leaving. The Gate crew had direct knowledge of who was in the community.

The Gateman's job was to assess each person coming in before allowing them into the community.

During the 1970s, up to 10,000 visitors a year would arrive at The Farm's gate.

When a new person arrived at the Gate, the first agreement was made. The Farm was definitely not a free hippie hotel with an open invitation to anybody showing up on its doorstep. If you were reasonably sane and interested in seeing what the place was like, you could visit for three days. If after three days you wanted to stay longer, it had to be because you wanted to take the next step toward becoming a member of The Farm.

Before new arrivals could go down on The Farm, they spent an hour or two, maybe more, at the Gatehouse with the crew. Depending on the time of day or season, there might be one person or a crew of four or five who got to know all new visitors a bit and checked them out. While most new arrivals were easy-going flower children, there was always a percentage of weirdos, sleezeballs, rip-offs and egomaniacs to filter. Sometimes just calling people out on their eccentricities and subtle manipulations would be enough to get them to straighten up and become normal, decent people again.

Following each piece of media exposure and publicity, more people would arrive. *National Geographic* published an article with a large photo spread. *Mother Earth News* printed several long articles. The news program *60 Minutes* profiled one of The Farm's members. The Farm was constantly in the news. The community also began sending Stephen and The Farm Band on tour, playing free concerts that included a slide show about the community. For the first ten years, between 100 and 200 new people would join The Farm each year.

In the early days, to join The Farm you had to make a personal connection with Stephen, acknowledging him as your spiritual teacher. It could be something as simple as going up to introduce yourself, stating your intention to make The Farm your home, not always so much in words but in a telepathic way that always ended with a hug.

By the mid-70s, The Farm population had grown to over 700, and in 1980 it was well over 1,000. Each year, over 10,000 people passed through the Gate as visitors, with up to half of those staying on for two to three days. Each person had to be integrated to learn their story and their intention and to make an agreement.

With so many people to process, it no longer became necessary or possible for every new person to meet directly with Stephen. Although people coming in still realized that the community was focused around Stephen's teachings, other than his weekly sermons on Sunday, his day-to-day role in the community fell more into the background.

The Gate had one man in charge of its operations, a gentle but firm ex-Marine named Leslie who had been with the group since San Francisco. He and a few others bore the responsibility for the Gate and security, with the authority to move someone from visitor status to "soaker." If you wanted to consider joining The Farm permanently, you were allowed to stay for up to four weeks to "soak in" what life there was all about and see if it was right for you. Soakers Meetings of 20 to 30 people would convene at the Gate every week. For the first meeting or two, new folks would simply be granted an extended stay. The mix always included single people and young couples, often with a new baby or a few kids. As they merged into The Farm's population, they would sooner or later get taken in by a household, find jobs by joining a work crew and become absorbed into the fabric of the community.

The Gate and the Law

Perhaps one of the most important aspects of the Gate and its role for The Farm was its interface with local law enforcement. Because The Farm and all of the roads and residences within it were on private property, police needed a warrant to enter the premises. However, rather than creating a point of contention, in certain ways, this suited the local sheriff and his deputies just fine. As in all agencies of the State engaged in this line of work, the sheriff serving Lewis County had his hands full dealing with local bad boys and ordinary crime. There was no need to meddle in the affairs of several hundred peaceful hippies camped in the woods.

It was at this time that some of The Farm's earliest important decisions began to bear fruit. The most important of these was a dedication to nonviolence. No weapons were allowed in the community. This eliminated virtually any residual fear from the neighbors as well as local law enforcement and sent out a clear message that The Farm was not a threat. Stephen regularly noted that "if we had been hippies with guns, the community would not have survived." Think Waco. Even Christians with guns can run into trouble eventually.

Another key element was the dedication to honesty and truth. This meant if someone representing The Farm spoke to the sheriff or one of his deputies, their words could be accepted at face value. This played out in a very real way around runaways and fugitives.

Thousands of kids ran away from home in the '70s, and a fair number would find their way to The Farm. Upon their arrival, The Farm Gate made each runaway take what might seem to be a surprising step, especially to the kid arriving at a hippie commune. The runaway was told to immediately call their folks and let them know they were doing OK and where they were. The upside of this was that any time local law enforcement was on the lookout for a runaway, they could simply drive up to the Gate and ask if the kid had arrived. At times, The Farm actually became partners with the police, made aware that a runaway was headed in its direction and asked to keep an eye out for them. This began happening so frequently that The

Farm soon established a residence specifically for runaway teenagers supervised by "house parents," a married couple and their kids. This Bunk House was actually built by some of the first runaway teenagers to arrive, guided by some more experienced Farm carpenters, and slept 40 or more.

In order to put the sheriff at ease, The Farm made an agreement. It would not harbor any fugitives from the law. Because members made this decision and stuck to it, the police did not need to gain a search warrant to enter The Farm and look for a fugitive, something that would have been an impossible task anyway.

To follow through on this agreement, when The Farm was initially founded, several draft dodgers came forward and turned themselves in. Arrangements were made for these guys to make amends by putting in community service at a nearby hospital. These bearded, long-haired ambassadors became high-profile representatives very much in the public eye as they went about their daily routines there. One mowed the lawn. Another worked as an orderly, his waist-length ponytail and piercing eyes leaving no doubt he was from the hippie Farm. For decades afterward, local Tennesseans would ask about the pacifists from the hospital who left a good first impression.

The Changeover

By the early '80s, things began to change. The hippie kids of the '70s were starting to hit their early 30s, when life gets more serious and it's time to settle down and have a real job. Visitor traffic had slowed down to just a thousand or so annually. Nevertheless, there was a steady flow of visitors, and the Gate continued to serve its purpose as the first interface with the outside world, establishing the perimeter that controlled visitor traffic, as well as preventing random gawkers and inebriated yahoos from freely cruising down its roads.

However, with the advent of The Changeover, the Gate had to be considered in an entirely new way. First and foremost was the money needed to pay people to take shifts at the Gate, which up to then had still been staffed 24 hours a day, 7 days a week. On top of the general staff, Leslie also asked for an annual salary for organizing, managing the Gate and security crew and being on call. In his role as The Farm's security, he also had a vehicle that would require fuel and maintenance. As the community began to formulate a budget and determine its true operating costs, the amount requested for the Gate loomed as one of the largest numbers.

Leslie asked for $25,000 in salary and to cover his vehicle, not unreasonable but no small amount, especially when considered today factoring in inflation from '80s dollars to now. A sizable number of people in the community felt it was time to abandon the Gate altogether and open the road to whatever traffic might come through, just like any other neighborhood in Tennessee. However, most members were not ready to take that step, highly valuing the privacy and sense of security that allowed women and children to walk The Farm roads freely and without fear, even after dark. Maintaining the Gate was deemed essential to holding on to the community's cohesive integrity.

Ultimately, it came down to a vote and counter proposals. A woman from the community offered to move her small business to one of the rooms in the Gatehouse, establishing a daily presence there. In addition, she would answer the phone and deal with any visitors that might arrive. Not only that, she offered to purchase a small motorcycle that she would use to monitor The Farm's borders. In compensation she asked for $8,000 a year.

At the same time, two members formerly from The Farm's electrical crew submitted a proposal to install a motorized gate that would automatically close every night at 7 PM. Each member would be supplied with a wireless beeper or remote control with a special code that would open the Gate after hours. A system would also be installed that would automatically open the Gate as anyone drove up to leave The Farm. This installation would have a onetime cost $8,000, including parts and labor.

It was hard to compete with those numbers, and Leslie lost the vote to the less expensive proposals. He became another victim of The Changeover, a loyal servant who had toiled in a unique position for the community for many years, now forced to figure out a new way to support his family.

After a few years, the woman who initiated the new Gate proposal passed the role on to someone else, and the position changed hands a few times until someone came along ready to continue for the long haul.

The job description of the Gatekeeper still includes having a working relationship with local law enforcement and being on a first-name basis with the immediate neighbors. Serving as the public face of The Farm requires learning how to integrate visitors of all types, from local Tennesseans to new-age nomads and potential new members.

The Welcome Center

With The Changeover now decades in the past, The Farm realized it needed an attitude adjustment, represented by changing the name of the Gate to The Welcome Center. In the early days, the Gate had been about establishing a presence, an identity, and setting bound-

New people interested in becoming members are encouraged to attend The Farm Experience Weekend in order to gain a better understanding of how the current community functions and operates.

aries, letting the hippie masses that arrived know The Farm was not a crash pad but a community. Stephen was fond of saying that "the gate swings out easier than it swings in," to dispel any notion that The Farm was a cult that held people against their will. It also inferred there were ground rules of expected behavior and that to live on The Farm you had to make a commitment.

Now as The Welcome Center, the purpose is to put forward a friendly face, an invitation to those arriving that they are welcome to visit, even stay for a night or a few at The Welcome Center campground, the hostel at the Ecovillage Training Center or in one of the homes that offer rooms to overnight guests. Visitors enhance The Farm's economy by shopping at The Farm Store and attending workshops, and they are able to participate in activities that include fundraisers, concerts, hikes and school events. Often even a short visit is enough to form a strong connection that continues after people leave. They may choose to support one of The Farm's non-profits and their many projects, or assist Farm businesses, from working with a Farm contractor to becoming a steady customer for products in The Mail Order Catalog.

Becoming a Member

It could be said the process for joining The Farm, becoming a member, has changed greatly, yet in other ways, it continues to share many aspects of how things have been since the very beginning. Someone wishing to join the community starts out as a visitor and then transitions to Resident, a status that signifies an agreement has been made, validating that the person or family desires to live in the community and test the waters. This is much like the period of soaking, only instead of a few weeks the Resident status could last two years or more. Just as before, the important thing is that new people are able to mesh with the community, and that it feels right for all concerned. In that way, it is still very much about the vibes.

Just as cities and towns around the country and the world have their individual feel and personality, The Farm has its own unique

New people interested in becoming members are encouraged to attend The Farm Experience Weekend in order to gain a better understanding of how the current community functions and operates.

flavor. Its roots remain firmly planted in the founding hippie generation, now blended with the slower, easy pace of rural life in the South, compounded by a global socio-political awareness similar to the broader awareness one might expect in cities like San Francisco or New York, but not present in the rural South. There is also an understated spirituality that is a reflection of core values. To the newcomer, these can be almost invisible because of their subtle nature, until surfacing overtly when triggered by some event or social interplay within the community. It may take a few years for a new person to encounter enough expressions of Farm values for them to gain a clearer picture of the strengths and weaknesses the community embodies, the gestalt vision and the challenges it faces in realizing its greater potential.

During those same years, the Farm's members have the necessary time to see a prospective member's polished exterior fade, a chance to encounter the individual's true personality and their strengths and personality faults. Over the first months or years of the initial honeymoon, new people are typically all aglow. They are so thrilled to finally be achieving their dream of shared life in community that a smile may never leave their face. After a year or two, old habits may emerge that were not evident at first. A lot can be revealed when the new person encounters their first controversy, or bumps heads with another member over some issue.

Of course, this works both ways. It can take just as long for the new person to accept that The Farm must deal with the same human issues that people have to contend with everywhere. Because we are together in community 24/7, week after month after year after year, there can be a depth and intensity to social dramas that people living in the isolation of modern suburban life never have to contend with. The prospective new member will have to decide if they still love The Farm after they have seen its faults.

All of this combined is tied into the membership process.

The very first step is to visit in person. So often, people will initiate contact, stating emphatically, "I want to move to The Farm," based on what they read on the website or in a book. Until a person actually visits, sees the place, meets the people and gains a feel for The Farm based on firsthand experience, it's all just a mental picture. Anyone interested in moving to the community is encouraged to visit, as many times as possible, before deciding to move to The Farm. A visit might last a few days or a few weeks. Over the course of many visits, the new person can start to see some familiar faces and hopefully make some friends.

Residents moving toward membership must have a sponsor, a long-term member who can help explain the steps, guide them through the hurdles and mediate if difficulties arise. The sponsor will be at their side when meeting with the Membership Committee, and maintain this role through the entire process until they are voted in and become members, letting the sponsor switch to the role of very good friend.

It could be said that the ideal new person coming to live at The Farm would be someone young, with the energy and enthusiasm to carry the community into the future. At the same time, The Farm does not discriminate by age. Each applicant is regarded individually. It comes down to both logistical issues and personality traits. Can you make it here economically? Do you carry a massive debt load? Where will you live? Can you afford to rent a home on The Farm while going through the membership process? How is your health? These are just a few of the issues that are considered with each person.

After some time, and the individual or couple's intent is clear, the community's Membership Committee will vote to upgrade their status to Provisional Membership. Provisional Members are announcing their intention to become full members. To a large degree, the Provisional Member goes about life in the community with the same rights and responsibilities as any other long-time member. Provisional Members are encouraged to attend meetings and fully participate in discussions and ongoing community activities. The two main exceptions: A Provisional Member does not vote in community elections for governing bodies like the Board of Directors or Membership Committee, and a Provisional Member does not vote to approve someone as a full member.

The Provisional Member and Housing

Resident cannot build or buy a home. Therefore, they must rent a place to live, and that can require patience and perseverance. Most rentals inside The Farm exist because a member is away for some period of time. The member may have left to care for aging parents, for reasons related to employment or for some other personal reason. Rentals are generally in limited supply and at times are in high demand. Often members prefer not to rent or rent only to guests visiting for a week or a weekend. Most places do not stay vacant for any length of time. Part of the process of moving to The Farm is getting in the flow and in the know, with assistance from the sponsor for a heads-up when a place is about to become available.

The Provisional Member has one key important privilege over the resident, the ability to build or purchase a house. Homeownership is very much about putting down roots. It is a powerful step symbolizing the blending of your future and the future of the community.

Because banks do not provide conventional mortgages, Provisional Members must be able to self-finance the purchase of a house or construction of a new one. Therefore, before a provisional member makes the huge personal investment of money and resources to acquire a home, they need to be very, very sure about their decision. If a few years down the road, things do not work out and they change their mind, The Farm will not buy out the person's equity in their home. They must wait until a willing buyer is found, another member or provisional member, however long it takes.

Not surprisingly, this aspect of joining the community reflects the complexity of modern life and contrasts sharply with the period of innocence embodied during the community's formative years in the '70s. Back in the day, people in their 20s or early 30s could show up with little more than a backpack and in short order find their way into one of the communal homes or tents. Today the majority of members have lived in the community for over 30 years and have seen it go through many stages, from bus to tent to shack, overcrowded communal living, poverty, working for decades to create the quality of life they have today.

At the same time, most new people coming in are unwilling to start from scratch, living as pioneers without running water or refrigeration. Both long-time Farm members and new people realize that such amenities are not luxuries. These represent a standard of living necessary to maintain the health and well-being of everyone in the community. Without a doubt, housing continues to be one of the greatest hurdles for anyone seeking to live at The Farm.

The Provisional Member and Employment

Unlike during the communal period of the '70s, the community no longer supplies the new resident an unpaid job in exchange for room and board. They must find a way to generate an income, to cover their own expenses and to contribute to the operating costs for the community, something that is the shared responsibility of every member and resident. This area of rural Tennessee remains isolated from the jobs and employment opportunities to be found in big cities.

Finding a position inside The Farm takes time. Most Farm businesses need to see if the new person is going to follow through on becoming a member and be around for years, justifying the time to train them and ensure their value to the company. These companies have very little turnover. Employment inside The Farm usually is based on skills, and there are a limited number of unskilled starter positions. People in construction trades willing to work on a crew outside the community can sometimes have an easier time finding a job right away. Often, the people who have the easiest time moving to The Farm are those with computer skills that are able to telecommute and maintain a position with an employer or company located elsewhere.

Fortunately, there is also a new generation of young people who share similar values from the hippie era seeking a quality of life in sync with their higher ideals of eco-sustainability. The Farm is frequently able to absorb and create a space for these young idealists who may indeed be interested or willing to live communally and with less financial security in order to pursue a life more in line with their personal philosophies. This next wave of visionaries is essential if The Farm is to continue beyond the founding generation and remain a viable example and model of alternative lifestyle for the world.

For As Long As We Shall Live

For the new person coming into the community, the membership process can seem overly complex and arbitrary. There is no question that it is very long and drawn-out. It is important to see that becoming a member of The Farm is much closer to starting a marriage than it is about simply moving to a nice community. Consider the basic marriage vows:

For better or for worse.
For richer or poorer.
In sickness and in health.
Till death do us part.
All literal, not symbolic.

Living on The Farm is about developing relationships that are deeply intertwined. Your friends are with you to celebrate your joys, but also to give you comfort and support in times of pain and grief, the ones at your side to care for you when illness strikes. Life in community is about sharing fortunes for as far as the road will go, the journey to the end.

As a member, you become involved in each other's daily life as family... all types of family.

Experience has shown that when someone becomes part of the community, they also bring their karma. They join not just as one person, but carry with them a lineage of extended family, which at any time during a member's life in the community may also need to be taken in. This can include the aged parent, the homeless dysfunctional brother or sister, the addicted child, the ones who are most likely to need care. We are our brother's keeper, in sickness and in health.

Every person who comes into the community has the ability to contribute to, or disrupt the fabric of the community. Often over the course of someone's time here, they may bring both.

Being in community is about sharing deeply meshed lives, much more than people may realize or are often ready for. There can be an intensity and depth, adding a richness that goes beyond the thin layer of existence found in the so-called normal life of mainstream society. At the same time, there is considerably less privacy. Everyone knows everyone. On the outside, you can have one face or personality you exhibit at work, a different one for friends, still another front when at church or as a member of a club. On The Farm, all of these different facets overlap, people get to see and know all sides. Life in community is not for everyone.

By remaining mindful, giving love the chance to grow for each new person, to become family, they are accepted for all that they bring, for better or for worse, for as long as we all shall live. When everything is working right, community brings out the very best a person has to give.

CHAPTER 7

Community, Conflict and Compassion

The Farm's ability to survive and continue over decades of change is due in no small part to the community's ability to resolve and endure conflicts large and small. A basic straightforward approach to resolving conflict has allowed the group to stay together, tempered with forgiveness and the recognition that time does indeed heal wounds, but that scars that go untended can fester and rise again to the surface. The returning conflict represents a new opportunity for those involved to resolve their differences.

Relationships

The Farm is unique for the depth of long-term friendships between members of the community. The majority of people living on the land have been there over 30 years starting originally as Founding Members or as children who grew up and are now adults in the community. There is a lot of family. These bonds form the glue that holds the community together.

At the same time, the ability to develop new friendships with people who have joined as members in more recent years provides the polish and luster that keeps life here fresh and renewed. What allows The Farm's newer folks to meld seamlessly into the fabric of the community is directly related to the community's ability to pass on the skills of conflict resolution as a tool for personal change.

It is important to accept that conflicts will arise. Personalities will clash. Buttons will be pushed. Toes will be stepped on. Not everyone is warm and fuzzy all the time. People express themselves differently from culture to culture. Living in community brings people closer together, increasing the chances for disagreements to arise. At some point, new arrivals may find themselves in a personality conflict surrounding an "issue."

It is also clear that even long-term relationships can become crossways, and may fall into repeated energy patterns. This explains why, even among close friends, sometimes it becomes necessary to admit that they have let petty disagreements push them apart. Each may feel or believe they can see how the actions of the other produced the wedge at the core of the disagreement. The real change only comes about when we reflect to see our part and what actions and changes we can make in ourselves. By assuming a person's goodwill, you can enter into the conversation together as two people wanting to understand each other and to be heard.

Most issues can be cleared up as simple misunderstandings that end with a hug. But the cycle of life includes larger dramas. Situations will arise—some serious, some not so serious—but they will take some working out. Learning how to talk to each other is a valuable skill.

The Path of Truth

The primary element that resonates to the core of The Farm's spiritual foundation is the profound respect for truth itself as a concept and practice. In a community formed by individuals coming together to seek spiritual understanding and awareness, it is assumed that each person can hold on to truth as a lamp illuminating the darkness of illusion, leading to greater clarity and sanity. Truth is seen to represent honesty in all relationships and exchanges.

Throughout the course of a day, all activity could stop in order to "sort out the vibes," the process of settling disputes and resolving differences.

From the Beginning

As the members of The Farm learned how to live together in community, the adherence to truth served as the touchstone for conduct in day-to-day living. Each individual was regarded as God's eyes into the world, carrying one piece of the truth as the Observer. The power of truth could be utilized to raise the collective intelligence for decision making and for resolving differences, rallying each person's input until the actual truth of the moment could be determined. In practical application, this meant that everyone was expected to express their feelings and put into words their observation and viewpoint of the truth.

Communal Living—Sorting It Out

For the first 13 years of The Farm's existence, families and single folks would live together in communal households of around 15 to 45 people. Residing in such close proximity to one another each day meant that people were exposed to each other's actions and habits in a way that doesn't happen when living separately. The subtle energy exchanges between a husband and wife were no longer hidden away but instead played out on a stage before an audience of observers. The sharp words and subtle intimidation that often take place between spouses unnoticed behind closed doors became obvious and unacceptable tactics when acted out before fellow housemates. Often, it simply came down to telling guys to being nicer to their ladies.

Encounters in the Workplace

Work crews provided another opportunity for people to bump up against each other and "work it out." The hierarchies and pecking orders found in usual workplaces were replaced by a system in which (ideally) no one had greater social position than another. All were considered equal. This meant that when someone felt they were treated unfairly, it was within their right to call attention to what happened and seek redress to their situation. If a person felt belittled, intimidated or bossed around, they were expected to speak up for themselves and not let the exchange go down.

When unbalanced energy exchanges are able to occur without being acknowledged and corrected, they linger in the subconscious mind. In order to clear the air, it became each person's duty to bring these actions to light, speaking the truth during what could be long sessions to "sort out the vibes."

Because each person agreed that their presence in the community was to pursue a spiritual path, this also meant that they were seeking this type of information about themselves in order to bring about personal change... at least in theory.

Each person's ability to hear the information and accept it played out in different ways. It is always easier to see the faults in others than it is to acknowledge these same faults in yourself. When delivered with love and compassion and coming from friends that you knew and trusted, the person receiving feedback would be able to hear and take in what was being said and phenomenal changes in personality could take place. Old habits could be broken.

Joel, a member since the early '70s, describes it this way:

> We discovered that a person's subconscious habits are only subconscious to him or her. It's conscious to everyone else. So we agreed that it is in a person's best interest to tell them about their bad energy habits, if it was necessary, and it must be done in a compassionate way. [It worked when] the person [who] received the information [understood] that it was given to be helpful, and looked inward to "de-condition" themselves from the bad habit.

However, when a person was backed into a corner, surrounded and outnumbered by people unhappy with their actions, walls of defenses could also be thrown up, blocking the individual's ability to take in the information and make positive steps toward change. We recognized that, in these often very sensitive situations, the element of compassion is paramount. Joel adds:

> Jokes and sarcasm, cynicism, criticism, competitiveness, one-upmanship are seen as power and control techniques. We can discuss differences without being feisty with each other, and in a way that doesn't demean the person we disagree with. We can all be together without any fences, [but only] if we each become aware of how we are feeling when communicating. We need to communicate from a reflective state and not be so attached to our positions that we communicate while angry or vengeful.

During the early days of The Farm, sort sessions could last for hours, and at times, all work would stop until the individuals involved could get straight with each other. Not every encounter worked out smoothly or came to a positive resolution, but enough did, so that learning how to mediate differences became a core element of the community's survival.

The More Things Change, the More They Stay the Same

Jumping ahead 40 years, the basic elements of working it out remain, but many things have changed. Farm families no longer live together in large communal households. Usually people live in single-family dwellings. Nevertheless, most people residing on The Farm were members during the communal years and have spent decades utilizing the process of speaking truthfully and directly to each other as a means of resolving differences. Approximately a quarter of the current community is next generation, young people who were born on The Farm that have absorbed these techniques through direct observation and osmosis.

Mark, a next-generation member born on The Farm, now married with a child of his own, recalls:

> There was a lot of working it out on the old Farm. We saw it through example with our folks and different people in the community working it out with each other. Whenever we'd [as children] get into trouble, we would be working it out with our parents or whoever we had transgressed against. [Now] I see that skill in myself and my friends, being able to communicate and work with people out in society on a greater scale, and it is a great strength we all bring.

Newer residents will, at some point, find themselves drawn into this means of conflict resolution either as observers or as direct participants in a disagreement. It is the way of The Farm.

"You have to be considerate of more people's opinions than you normally would have to deal with," explains Susan, a 35-year-plus member.

> People care about how we treat each other. So when issues come up you have to take the time to listen and hear people

out and express your opinion and let there be quite a few interactions. You don't get immediate gratification most of the time. You must have patience and a willingness to go through a process.

Step by Step

Pamela, who has been with the group since the days in San Francisco, says, "Talking to each other is primary. You learn not to just talk around somebody or get mad at them but actually talk to them."

In most situations, the breakdown of communication is between two people. Although it can be difficult, the very first step is to talk directly to each other, one-on-one, being careful to listen and take in what the other person is saying, as well as putting out your side of the argument or feelings in the situation. More often than not, this is enough. Failing to talk, avoiding this difficult conversation, can allow the gravity of the situation to get blown out of proportion. Assumptions about the other person can melt away as you come together face-to-face and recognize each other as human beings that are inherently good, who make mistakes, and remember that you are actually friends.

A Fair Witness—The Path of Mediation

Sometimes the wedge between people is too wide for them to resolve without help. Each may feel distrustful of the other or have experienced a hurt so painful that listening and seeing the other person's viewpoint is unachievable without help. To facilitate resolution, instead of getting together one-on-one, each person brings along a fair witness. Ideally the two people chosen for this role are accepted as honest and respected for being fair by both sides. It works best when the fair witness is also an effective communicator, since another aspect of their role can be to speak on behalf of one party or the other. In situations where conversations rise in intensity and one or both people engaged in the dispute lose their cool and get defensive or uptight, the fair witnesses can interject and call on both sides to take a breath, calm down and remember why they have come together, ultimately to resolve those differences.

Going one step further, it may be necessary to have a more formal mediation and call on the service of a trained mediator. The mediator is a person present that does not speak for or stand with either party, someone totally outside of the issue, allowing them to be completely neutral. Many people in The Farm community have received training as mediators and have assisted in the local county criminal justice system, helping aggrieved parties achieve closure and restitution, a process that works especially well with cases involving juvenile offenders. A member of The Farm worked on legislation to enact a state law offering mediation as an option for juvenile offenders, making the services available in every county of the state.

Crossing the Line

On the early Farm, with so many people living in such tight quarters and with the acknowledged agreement that it was OK to "get up in each other's thing," egos were kept in check and for the most part no one got too far out there.

When the community faced an economic collapse in the early '80s, for a time all rules and agreements became somewhat up for grabs and in a way redefined. As the population shrank from the peak of over 1,000 people down to around 100 adults, there was no longer the intense need to work it out, as everyone became more focused on their own families, raising their kids, establishing careers and taking care of their personal lives. Members experienced more individual freedom, and there was no sense of compulsion to monitor or engage in the affairs of a neighbor or friend.

Unfortunately, over time this also means that occasionally someone will cross the line and allow themselves the luxury of anger, which in their mind becomes personally justified because of some rationalization. In this case, we are talking about more than a heated discussion, where the situation feels on the edge of physical violence or the verbal tone is extended and abusive. It happens. People act out, lose it. Something goes down. Those observing the conflict may feel strongly enough about what went down to call for a Disciplinary Inquiry.

The Disciplinary Inquiry is part of The Farm community's bylaws. The Membership Committee goes on a fact-finding mission, seeking the truth, talking to those directly involved, to casual observers, anyone who might have a piece of the puzzle. The Membership Committee tries to ascertain the why, why did these actions happen? Their task is not only to consider the immediate cause, but to question what may be beneath the surface, the underlying reasons or root cause. More often than not, actions in the present will stem from something unresolved from the past.

Nonviolent Communication

Nonviolent Communication, or NVC, is a national movement that has found its way to The Farm as a method of improving communication skills and mediation efforts. It embodies many of The Farm's

Nonviolence remains one of The Farm's core values. Above, Farm members and former members come together for a meeting a few weeks after 9-11 to found the nonprofit PeaceRoots Alliance.

techniques for "working it out," but helps refine and improve outcomes by consciously identifying methods of conflict resolution. Paramount is the art of deep listening. Often in an ego confrontation or argument, we are only half listening, instead lost in our heads, and busy planning a rebuttal. We're not taking in the information, and letting the other person know that we have stopped and truly heard what they had to say.

A simple exercise of Nonviolent Communication is to repeat the person's point, which can be even better if restating in your own words, not just parroting their statement, indicating that you actually understand what they have said. This act requires you to really listen and consider their situation, which may even modify your own feelings and soften your position, before jumping into your own side of the disagreement. When someone feels truly heard, they are better able to let go and be open to the other person's position as well. By leaving a few moments before the next person speaks, they will be listened to with the same respect.

Forgiveness

Ultimately, the only way anything is ever resolved is through forgiveness. The person committing the offensive behavior must truly understand the ramifications of their actions, exhibit remorse and seek forgiveness. It is then up to those on the other side to accept the sincere apology, offer forgiveness and provide the space for change to take place.

Space

Although forgiveness is the first step in healing, there are usually emotional scars. While people engaged in a conflict may come to a resolution, it doesn't necessarily mean they immediately become fast friends. That only happens in the movies. In real life, sometimes things may work out best if people are allowed to have space and go their separate ways.

This is one of the great advantages The Farm has over many other intentional communities. In smaller communities of 10 to 20 people, any discord affects the entire group dramatically, even leading to the breakup of the community. In contrast, The Farm's population now hovers between 160 and 200 people. The natural tendency is for members to have a tight circle of a few close friends, a larger circle that they might resonate with in a variety of ways, with overlapping circles that take in most everyone else, including a few people they may choose not to be around. Because the residential areas of The Farm are very spread out, and with multiple layers of people to hang out with, the community is able to absorb a lot of personal drama or trauma and allow people to get on with their lives. Social dramas usually only directly involve a small number of people, leaving the vast majority as mere observers or completely oblivious. It pays to remember that everyone is more absorbed in their own movie, the one in which they are the star. While your personal drama may feel like it stands out like a wart on the end of your nose, for those not directly involved, the incident may be no more than a blip on their radar screen. This degree of separation allows the community to roll on, absorbing the ebb and flow, seeing the cycles of tension and relief pass through almost like the seasons.

In the end, it comes back to friends. Our friends provide moral support when they see things from our point of view or tell us when we make mistakes. Our friends love us even when we blow it. They give us the room to change and possess the patience to wait for that change to happen.

When living in community, conflict is inevitable and compromise is essential. The Farm's ability to give and take is what makes the community truly sustainable.

CHAPTER 8

Tribal Cultures

THE DOMINANT CIVILIZATIONS and pop culture of the West have always maintained a nostalgic affinity for tribal cultures. Unfortunately, so often it seems our society is more attracted to the idealized symbols that native people represent while ignoring the actual and current struggle faced by all native people across the planet.

Tribal cultures embody a connection to natural forces and the Earth that is seemingly lost by the West. New-age movements often borrow heavily from Native American traditions, which represent a bridge to a saner relationship between nature and human existence.

Resonance with tribal cultures underlies much of The Farm's development, which could be described as the rebirth of tribalism in a Western European society. This connection to the values held by Indigenous people has been exemplified throughout The Farm's history and continues to flourish through real actions.

Beginning at Birth

The dedication to natural childbirth formed a bond not just between the couple and their newborn, the couple and their Midwives, but with the community as a whole. Birth became the link fusing the raw forces of nature with family and community, the essence of tribal relations.

The Farm's participation and support of The Longest Walk in 1978 thrust the community into the heart of US indigenous culture.

In the mid-70s, members of Awkwesasne, the nation of Native American Mohawk straddling the US and Canadian border in upstate New York, began communicating with The Farm, expressing a desire to receive training in midwifery in order to bring those skills back to their community. Several women and their families came down to live in Tennessee and begin a hands-on apprenticeship. At the same time, their husbands and a single man from Awkwesasne came down to live on The Farm and receive training as Emergency Medical Technicians, or EMTs, with The Farm's ambulance service.

In 1978 The Longest Walk, a gathering of Native American tribes from across the US, called attention to the economic conditions on reservations and broken treaties and raising awareness among the general population and even more specifically among Native American youth. The Longest Walk began on the West Coast with a couple of hundred participants intending to walk all the way to Washington, DC, picking up representatives from various tribes and Indian Nations as they passed through different states. As the Longest Walk converged on the many hundreds of small towns and cities, the presence of Native Americans generated press and media coverage that spread the message and, in its own way, helped put pressure on bureaucrats in Washington, DC managing Indian affairs.

The Longest Walk was loosely assembled by the American Indian Movement, or AIM, a group of Native Americans leaders. AIM's leaders were, to various degrees, in sync with the progressive leftist political movements of the time, and their presence played a vital role in the spirit of the event. Aggressive actions, such as the occupation of the former prison on Alcatraz Island in 1971 and the Bureau of Indian Affairs building in Washington, DC in 1972, brought media attention and helped galvanize the group but also made it the target of the FBI. The 1973 protest/occupation by AIM members of the monument to the 1890 massacre at Wounded Knee, South Dakota, resulted in a standoff shootout with the FBI lasting 71 days, ultimately culminating in the death of two AIM members before an agreement to disarm could be negotiated.

The Longest Walk was like nothing Americans had ever seen. Known to most people only through television westerns and John Wayne movies, the country's Native Americans were suddenly very visible as an ethnic subculture that had been submerged within the country, now cast into the light by the hundreds of Indians walking through Middle America. The Walk was a rallying call especially to "traditionalists," those whose lives symbolized a resistance to assimilation into Western culture in their communities. It provided an opportunity to showcase tribal identities beyond the powwows held in many regions with larger Native American populations.

The Longest Walk brought together Native Americans of all types and ages, from veterans of World War II and Viet Nam to aging grannies and multigenerational families. It appealed especially to the young, who were on the front lines of unemployment and hopelessness so prevalent on the reservations.

When The Longest Walk reached Pennsylvania, they were about 700 strong. Sensing the power and importance of this event, The Farm decided to send up an ambulance to help support the walkers and their mobile community. The Farm had one member who was Native American, a Muscogee Indian named Mark who would be in charge of the ambulance and crew, which included the EMT trainees from Awkwesasne and one of The Farm's Radio Crew to assist with communications.

The ambulance had a CB radio, and walkie-talkies were given to various representatives among the walkers and their support vehicles. This allowed them to call The Farm ambulance for any emergency. Fortunately, the Crew never had to deal with anything too serious, primarily heat exhaustion and blisters. The ambulance also had a police scanner, which made it possible to track the communications about the Walk by the authorities. The Walk was under heavy surveillance, with helicopters and police escorts a daily presence.

The march into DC took about six weeks, with the group setting up camp often for up to a week at various points along the way. Word was getting out, and the final encampment in DC had about 4,000 Indians and 1,000 white supporters. The Farm sent up a couple of dozen additional representatives, including Stephen and his family, to be part of the many rallies taking place across the city. Events included a concert featuring nationally famous Native American performing artists like Buffy Saint Marie, a benefit boxing match with Muhammad Ali, speeches on the steps of the Capitol and around the city. The energy was very high. It was the first time that Indian people had converged in numbers this large at the seat of power and the colonial government that had taken their lands and committed genocide on their people.

One afternoon, the march arrived in front of the FBI building for a press conference. A podium and public address system with a microphone were set up, and suddenly it became evident there was no place to plug into power. The FBI guys on the other side of the barred entrance just smiled. Fortunately, one of The Farm's support vehicles had its own generator, and we were able to supply the electricity needed by the PA. For once, the voice of the tribal people was not silenced by the government.

The opportunity to be involved with this unique assembly of Native American elders was a very eye-opening and transformational experience. At the same time, the close contact dissolved illusions, with the awareness that Native Americans carry the normal faults that all humans share, placing the problems within The Farm's social dynamics in perspective.

Land of the Maya

Without question, the most significant connection to tribal cultures by The Farm Community took place after its newly formed relief and development organization, Plenty, sent a team down to Guatemala after a devastating earthquake in 1976. What The Farm discovered was a centuries-old Mayan culture that was still very much intact, with the Indigenous Maya representing over 75 percent of the country's population. In many ways, it was like stepping back to a time when the relationship of people to the land and their tribe remained as the focal point for human existence. This extended direct contact with an ancient Indigenous culture had a dramatic effect on The Farm and provided a living example to the community, clearly illustrating how tribe and family are interconnected.

The work in Guatemala was the biggest project The Farm had ever taken on. A camp was established to function as a base of operations. Over the duration of the project, as many as 200 people were dispatched from The Farm to be involved in the relief work. Now, in addition to living the dream of creating a new utopian society, each person in the community could feel they were actually making a difference in the world and in the lives of others. Daily communications via the ham radio back to Tennessee gave everyone the sense that the entire Farm Community was united behind the effort.

In the months after the initial state of emergency had passed, the Plenty volunteers began to recognize that the earthquake was only the beginning of the problems the Mayan people of Guatemala faced on a daily basis. The work by the members of The Farm seemed to be just scratching the surface.

San Bartolo

For the first two years, Plenty volunteers worked in the town of San Andrés Itzapa, where downtrodden farmers served as virtual slaves on large plantations. When Plenty's service to the poor began to draw the ire of local government officials around Itzapa, the entire camp was moved to the state of Solola and a small village called San Bartolo, two hours deeper into the mountains.

Everything was different in Solola, where most of the population worked as independent farmers tending small plots scattered across the steep mountainsides. The vast majority of the Mayan population wore colorful handmade clothing woven on primitive backstrap looms, a form of weaving that had been handed down for centuries. Although still quite poor, the people had dignity and a sense of pride, a radiance that had an immediate appeal to the Plenty volunteers.

Solola was like a paradise on Earth. Located further up in the lush green highlands at 7,000 feet, the area lived up to the phrase "Land of Eternal Spring," with year-round temperatures never going above the mid-70s in summer or below 40 in winter. San Bartolo had a view out over Lake Atitlan, one of the largest and highest natural bodies of water in the world, formed by a volcanic crater. Five volcanoes surrounding the lake were visible from the camp, including an active one called Fuego (fire), which regularly spewed out plumes of smoke and ash. There was a magic about the place that was tangible, almost surreal.

The villagers of San Bartolo were just as curious about the hippies as the Plenty volunteers were drawn to the Mayan culture. The new setting was very much in the middle of the "aldea" or village. Plenty had leased a large home to serve as the new base of operations, and it came with a Mayan caretaker family living just a few feet from the back door of the kitchen. Other village homes were less than a stone's throw away. It was a very direct and deep immersion in the Mayan way of life. Teenage girls from the village enjoyed visiting the camp and cooking on the large restaurant-sized gas stove in the kitchen, teaching the Plenty people the art of making tortillas by hand. Women from the village began sharing their knowledge on how to weave on the backstrap looms. Real friendships developed.

Beyond what we were experiencing through our work on the projects, it was the day-to-day living amongst the people that brought home the real heart of tribal culture. Wrapped in the cloak of Christianity, there was an ancient appreciation of the divine that emphasized humility and gratitude. At the center of every home was the altar, four to five feet wide and a foot deep, dotted with burning candles illuminating a crammed odd assortment of figurines, objects with special meaning, trinkets and bottles of Coca-Cola. The altar was both mysterious and captivating, an everyday reminder of the spirit that underlies all things.

On holy days, the air was alive, awakened by the powerful smell and thick smoke of copal, an incense made from evergreen tree sap. The aroma carried with it the muttered prayers of grey-haired elders, perhaps a prayer for good harvest, the sacred corn that would fill the family's stomachs and allow them to survive another year. Corn was the rhythm of the seasons. It was the foundation, quite possibly the only food, at every meal. The seeds were planted, tended by the fathers and the sons, and the rains came and the sun shined each day. So much has been lost in our modern world, our tie to the cycles of life, the intimate knowledge of our sustenance, hardened by cement and blurred by speed. The memory of what we had forgotten was still breathing in the Maya.

Across the small valley below the Plenty encampment was a Mayan family compound, its borders outlined by dried cornstalks defining its dimensions. Around its perimeter were a number of small huts, walls made from the same stalks of corn, each covered by a palm thatched roof, bedrooms for the many families: the brothers and sisters, along with their spouses and several children, one for the patriarch and matriarch of the clan. The elders, no longer able to work in the fields, no longer able to carry the water, the firewood, but still part of the family circle, making their contribution, to tend the fire, to pat the tortillas, just as they had always done, together, the people of the corn.

Throughout each day, the women would sit together in the courtyard, to weave, to create the cloth, the very clothes on their backs, the clothes of the family, thread by thread, the wooden spindle passed back and forth, tamping row by row, an inch a day, perhaps two. The richness of color in the "tela," the cloth, was beyond beautiful, a textured warmth no machine-made clothes can ever come close to. Clothes made by hand, by dedication, by love.

It was a privilege to be served around the fire, the honored guest, a hot cup, "un café," a coffee roasted on that same hearth, of beans from bushes down the hill, a flavor smooth and rich, to share the smiles, the humor of getting to know each other.

In the developed world, we take so much for granted. We assume our water is free of parasites and that its supply is virtually unlimited. When water is in short supply and must be carried often a very large distance, one doesn't think about using it to wash hands. Flush toilets simply do not exist. Living side by side with the Maya, Plenty volunteers could experience firsthand the invisible oppression that the absence of clean water imposes on a people.

From 1978 to 1980, Plenty volunteers brought clean water to thousands living in villages across many parts of the country. Instead of simply putting a Band-Aid on the symptoms of poverty, Plenty's water projects in Guatemala truly improved the quality of life. But real help is not about the great white benefactors bestowing aid to the needy. There has to be a shared respect.

In conjunction with the water projects, Plenty volunteers would host meetings to educate Mayan villagers on the importance of clean water and the need for proper sanitation. A microscope was set up so that villagers could see slides full of wiggling parasites. The link between bugs in your belly, diarrhea and abdominal pain was made easy to understand. A healthy man is able to go to work and support

A Plenty volunteer stands with members of the Mayan community of San Bartolo, Guatemala to celebrate the first rush of water from the new village water system.

his family. Kids who are not sick are able to attend school. A people who are kept strong are more likely to hold on to their culture.

In the village of San Bartolo, there was a special night, a celebration near the end of Plenty's time there that seemed to capture the friendship and bond that had grown between the hippies and the Maya, two very different tribal cultures. Plenty and village volunteers had just completed work on a new school. All of the dignitaries, officials and politicians who had made their way to the village for the official dedication had gone.

The band played, a single marimba worked by three musicians, each with his part, the bass, the alto, the soprano, the full organic tones of wood vibrating through the air. There are no sad songs on the marimba. Small groups danced in several circles, holding hands. A woman would go to the center and spin around, reaching to take off a man's hat and put it on her head, spinning again, a twinkle in her eye. Then returning the hat and grabbing another, teasing, laughing, joyous, under the stars. Together, as one, the dancing went on into the wee hours of the night. No one wanted it to end.

The Political Climate Changes

Ever since the overthrow of the civilian elected government by a CIA-orchestrated coup in 1954, Guatemala had been under the rule of a long line of generals. After the Cuban revolution in 1959, the US military had increased its presence and influence in Latin America, suppressing any populist movements through harsh intimidation, eliminating (murdering) any union, student or religious leaders speaking out on behalf of the poor.

In a very real way, the earthquake of 1976 had upset the checkerboard and put the entire country in a state of chaos. Desperate for aid on all fronts, Guatemala had cracks in its infrastructure and control that groups working on behalf of the poor could squeeze through, and organizations like Plenty were able to establish a presence.

Life in Guatemala changed dramatically after Ronald Reagan became President of the United States. Even before he took office, it appeared as if deals were being made and the word was clear: use any and all means necessary to maintain control. For the Plenty volunteers, the idyllic life in the Land of Eternal Spring was to be overshadowed by a dark cloud.

The fear intensified with the emergence of the death squads known as the White Hand, named for the mark left on doors at the homes of the disappeared. In conjunction with the overt military campaign, the real counterinsurgency work was being done outside the realm of accountability. Day or night, armed men would arrive, murdering anyone even suspected of not fully supporting the government, their mutilated bodies dumped in the center of a village as an example of what would happen to those who questioned the military's authority. Entire villages were burned to the ground, turning any survivors into homeless refugees. Ravines with scores of dead bodies began to become commonplace, reported weekly in the country's newspapers with pictures and graphic detail.

One story in particular heard at the time demonstrated this brutality and sent a clear message about the intent of the oppression. In a town just across the lake from our location in Solola, a party was being held for two students who had just graduated and were returning to serve their community as doctors. Armed masked men arrived, took the two students away and murdered them. It was clear that the life of anyone helping the people was at risk.

Farm men with their long hair and beards fit the profile of the Marxist revolutionary. Plenty people dressed in the same handwoven clothing worn by the Mayans, a visible blatant statement of solidarity. Virtually all of the projects were in direct support of the country's Indigenous population, working with the poorest of the poor.

Everyone in the camp began to question how much longer the work could continue? There was an illusion of security, believing that it would cause too much controversy in the US media for any Americans to become targets. However, as stories circulated about American Catholic nuns and priests being tortured and murdered, even that barrier seemed to be dissolving.

In the late summer of 1980, the members of the camp, Plenty's directors back in the US and Stephen decided it was time to leave. The Farm's Greyhound bus arrived in September to transport everyone back to Tennessee. It was a very sad good-bye.

Where and whenever possible, Guatemalans with the necessary experience were hired to implement and complete ongoing projects. Over the coming decades, Plenty would continue to receive proposals and take on various projects in Guatemala, using the model of working with local organizers, rather than sending down teams of American volunteers. The Guatemala experience set the tone for Plenty's mission, to work in support of native peoples and their struggles.

Stateside

Back in the US, the participation on the Longest Walk established a number of contacts and led to long-term friendships in Native American circles, opening the doors to future Plenty projects. One of the first to get traction was at the Pine Ridge reservation in South Dakota.

The concept was simple: Help people take control of their lives and put their energy into something positive, a home garden. The request

For over twenty-five years, Plenty has worked with Lakota Sioux families across Pine Ridge Reservation in South Dakota to prepare and maintain gardens to augment their diets with fresh organic produce.

for funds and assistance came from a resident on the rez who would go to the home of anyone wanting a garden, till their ground and supply seeds and plants to get them started. The first year, there were 5, the next year 25 and, by the fifth year, 200 gardens throughout the reservation, and eventually grew to around 500. Plenty was able to be the conduit for funds from other non-profits that wanted to help empower Native peoples but did not have the direct connections to facilitate projects.

Over the last several decades, Plenty has continued to nurture this relationship with Pine Ridge, sending teams of carpenters and students from The Farm School to build a house, do home repairs around the reservation and develop the infrastructure for a camp where volunteers could stay during their time in South Dakota. The work has had multiple benefits, supplying a helping hand to those in need and a way to give volunteers the chance to connect with something deeper, the link to a resonance with the Earth and spirit that Indian people have carried for thousands of years.

Native Voices

The culmination of these collective experiences also made it possible for The Farm to serve as a microphone for the voices of Native people. In the late '70s, its Book Publishing Company began printing and distributing works by Indigenous authors. Over the decades, the number of these books expanded until the company dedicated an entire catalog to Native Voices, becoming one of the primary publishers for Native American authors in the world. Sales numbers are small, no million-sellers here, but these books serve an important niche audience and keep the avenues open for the words of Native people to remain alive and relevant.

The Farm as a Tribe

An estimated 5,000 people have lived and spent significant time in The Farm Community, most during the early communal years. The experience transformed people in a way that left them forever changed, a connection that persists as meaningful and real despite the passage of time. During the period leading up to The Changeover, more than 1,200 people lived on The Farm; the breakdown of the communal economy caused the dispersal of hundreds far and wide. But what has endured so many decades goes beyond friendship.

Dictionaries define the word "tribe" as "people with a common culture or character, a social structure that establishes links between families." The Farm went beyond the hippie subculture to become a

Once a year members of The Farm "Tribe" travel from all corners of the earth to gather together on the land and renew connections.

unique social experiment that created emotional bonds that remain alive in all who have been touched by their common connection. It spans multiple generations woven ever more tightly by the cross-pollination of families through second-generation marriages and subsequent grandchildren. In seeking ways to define these relationships, the word "tribe" comes closest. Understanding The Farm as a tribe takes in both the strengths and the dysfunctional weaknesses that exist within families and tribal communities, recognizing that the relationships go beyond random, even in their imperfection.

In that sense, The Farm Community is seen by its tribe as the reservation, the sacred land that symbolizes the greater whole. The land serves as a unifying vessel that holds the memories, the dreams and the energy of everything that has taken place within its boundaries.

And like the tribes of Indigenous people today throughout the world, the majority may never live on that land, yet they still are part of that common essence that links them to something larger than their immediate family. It also takes in the idea that there are people who share the same values and may be considered members of the tribe who have no immediate connection to The Farm, distant clans with their own set of experiences and circles of community.

As The Farm's communal period recedes farther into its own historical perspective and as its founding generation gradually passes on and departs from this plane of reality, the question remains open: What core values will persevere? Can this tribe survive the outside and internal pressures that challenge its existence, beyond the founding generation? Will the subsequent generations be able to maintain and forge new bonds that have kept the community intact throughout its history? Only time will tell.

CHAPTER 9

Following the Spiritual Path

FROM ITS VERY INCEPTION, The Farm was established as a "spiritual community" and registered with the state of Tennessee as The Farm Church. Stephen taught and the community recognized that intentional communities founded on spiritual principles were much more likely to survive beyond a few years, and there was history to back up that belief. Over the last 200 years of intentional communities in America, those based on political or social economic ideals had a life span of around 10 years, while those founded on spiritual principles typically endured 25 years or more. The spiritual awakening experienced by the hippie generation is what made The Farm happen, and why it exists today. Community is so much more than the buildings and the roads and the trees. It takes spiritual connection to endure.

Spiritual Teacher

Since Stephen Gaskin had already been a professor at San Francisco State, his conversion to a teacher of spirituality was not a giant leap. About a decade older than most of the young people massing in San Francisco, he was able to address the class with the learned air of authority, while calling on the natural maturity that comes from wider life experience. By assuming the role of spiritual teacher, he was in some measure stepping in to fill a void by developing an avenue of spiritual understanding that did not require assuming the doctrines of another culture, be it from the West or the East. In many ways, his role was to articulate the lessons learned through the application of universal principles, using language that could be understood and expressed in the context of modern life. At times, this meant interpreting or learning to recognize how the principles of Eastern philosophy were also expressed in our Western culture, understood in a new way through the spiritual awakening of the counterculture.

Back in San Francisco, the early free university classes conducted by Stephen were very much about an exploration of consciousness. Psychedelic "religious" experiences had exposed people to new realms of possibility, in touch with "the All" and a sense that the individual and the universe were one, connected by the very molecules of existence. People were left profoundly affected and filled with unanswered questions and began coming together to share insights and explore answers, turning to the wisdom as expressed through the sages of the world, be they old-world religions or more esoteric doctrines, such as astrology or new-age mysticism.

As in comparative religion courses in colleges and universities, Stephen and the people attending his Monday Night Class made an effort to examine spiritual and religious philosophies to determine which elements represented essential truths. By making this direct attempt to decipher and identify the common teachings of these spiritual doctrines, the participants in Stephen's classes began to recognize that core principles could be identified and utilized as a roadmap for the navigation of life.

In order to avoid the trappings of dogma and ritual, Stephen and The Farm intentionally decided not to give their outlook a name, calling it simply The Farm Church. For the most part nothing was written to define it, other than the transcriptions of Stephen's lectures, edited into books in which philosophical topics were explored and meaningful stories told.

After The Changeover, it became even more difficult to identify universal or commonly held beliefs by Farm members, in ways a reaction to the domination of a single voice enunciating spiritual morals to be adopted by the community as a whole. Individuals were and continue to be free to define spirituality and religion for themselves. Today, members of The Farm embrace a full range of expression, with members that fall somewhere in a spectrum that includes various levels of practice within established religions, nature pagans and tree huggers, proponents of New Age woo woo, along with acknowledged agnostics and atheists. It would be more accurate to say The Farm's spirituality is defined by the common values that have been woven into the fabric of The Farm since the very beginning and that still run true today.

The concepts presented here are not new. Many have been repeated throughout the ages, in every language, by philosophers and shamans, preachers, teachers, rabbis and clerics.

The spiritual path is defined by moral principles which can help guide us through life, directly affecting every choice we make.

Spirituality and Religion

To begin, it can be helpful to understand the difference between religion and spirituality. World religions are typically based on events that took place hundreds, even thousands, of years in the past, rewritten over and over again, each shifting to match the language and attitudes of the time. In a quest to provide answers and explain the unknowable, religions ask adherents to put forth a faith in the unknown and to follow that faith blindly without question.

In a generation coming of age in a wave of planetary consciousness, there was a general respect for core concepts of established religions like Christianity and Judaism, but the historical and modern-day hypocrisies and evident shortcomings made it hard to accept these religions as anything other than ancient doctrines practiced through dogmatic rituals, with many of the values watered down or talked about but not applied. The truths they carried were undeniable, more like pieces of a puzzle, needing one from every religion to see the complete picture.

Spiritual teachers representing Eastern philosophies had a historical reputation as keepers of knowledge, with gurus that could provide answers and a path that followers could adopt almost like a recipe, delivering the desired result of a spiritual life. This was propelled into Western pop culture or mainstream awareness when The Beatles become involved with the Maharishi, whose transcendental meditation offered a secret mantra that one could repeat silently in order to achieve higher consciousness. Young people in large numbers began to align themselves with an array of spiritual teachers, assuming the role of a devotee. The stereotypical Indian guru bade their followers to adopt the appearance of a yogi by dressing in white, wearing turbans, further changing their persona by assigning each disciple an Indian sounding name. Although Eastern religions offered new avenues of thought and a path to spirituality, they too came with extra baggage, cultural superstitions along with beliefs and rituals that had little relevance.

In contrast, spirituality can be considered a framework of moral principles that can be used as a foundation for decision making and a guide for personal conduct. When faced with life's difficulties, these morals offer direction for the individual every step of the way. By acknowledging and putting into practice core spiritual values into everyday life, each person is carried toward their ultimate goal, a sane existence filled with happiness, one that someday ends with a sense of fulfillment.

Religions typically ask their followers to base their faith on supernatural events that took place in the past, using these as evidence to demonstrate the existence of a higher power. Those who follow a spiritual path are able to use their own direct experiences to influence and guide their life choices and to gain a sense of universal oneness connecting everything. Pivotal moments happen in every person's existence if we are there to notice, opportunities to gain wisdom and appreciation.

In the search for greater understanding, Stephen and participants in the class began to acknowledge that all religions share essential truths. The differences we see emanate from variations in cultures and their place in time. By working to identify, follow and apply universal moral principles, the group began to move toward a way of life that could express these ideals through day-to-day actions.

The path charted by The Farm accepts that we can honor and respect all religions for the truths and teachings they hold and their cultural connections rooted in families and traditions. Religious celebrations function as symbols representing our link to something greater on many levels, steeped in the communion of shared experience. When we participate in these ceremonies and traditions, we connect again with something greater than ourselves, be it a shared traditional history or the life transformations that these celebrations represent.

It is no coincidence that many religious holidays coincide with cycles in the natural world, drawn from pagan and nature-based religious origins. One of the easiest to recognize is the celebration of Easter, taking place on or near the spring equinox. The word itself comes from the name of a Nordic goddess "Eostre" or "Eastre" who represented fertility, of critical importance to ancient people whose survival depended upon abundant crops and the bounties of nature. Symbols like the Easter basket and eggs are tied to fertility rituals performed by farming peoples of Europe who would collect and place the nests of plovers, a ground-nesting bird, in their fields as a prayer for bountiful harvests. Artifacts dating back over 25,000 years combine the head of a rabbit with the female form or goddess, both symbols of fertility. Our ability to recognize cultural celebrations and traditions as expression of humanity's relationship with the natural forces helps to broaden our understanding of universal principles.

In one of his later interviews, Stephen Gaskin expressed it this way, "All religions are built from the wiring diagram of the mind of man. It should be no surprise that they all came out the same."

Sacraments

Sacraments such as the wine and bread of communion are symbols used to help express religious concepts. The sacraments expressed in many practiced religions often stem from events that took place hundreds or thousands of years ago, kept alive through stories and ritual.

By seeking a path of direct experience, the Church of The Farm Community recognized the sacraments that are present in the here and now, important life transformations that have the ability to touch each person and impart to them a sense of the profound.

The Sacrament of Birth

One of the most powerful experiences that came to impact the hundreds of people coming in contact with The Farm was that of birth, a practice enshrined in the phrase "spiritual midwifery." Over the last 100 years, much of the Western world has had the miracle of birth removed from the realm of direct experience. By the late 1960s and '70s, shuttered behind the doors of operating rooms in clinical sterile environments, birth was no longer a natural process, but a medical procedure. Fathers were completely removed from one of the most significant and direct life-changing moments one can ever encounter. Drugs made mothers unconscious and unaware of their own personal miracle.

Back in San Francisco, several of the women associated with Monday Night Class had become mothers and were unhappy with how they had been treated in the hospital. Stories also began to circulate about a few brave women that had chosen to give birth at home. When The Caravan left to travel across the country, among the group were several women, including Stephen's partner Ina May, who were due to give birth during the journey. Several babies were born in buses along the way, and the impact of those experiences was so powerful that it affected not just those present at the birth, but everyone in the group. After they arrived in Tennessee, the sanctity of birth was acknowledged as one of life's most important sacraments and a cornerstone of The Farm Church, for the power it has to change lives forever.

The Sacrament of Death

One of the babies born on The Caravan did not survive. Soon after arriving in Tennessee, a young man was killed by a lightning strike. These events and others to follow forced the community to face the entire circle of life, bringing Farm members into direct contact with another universal sacrament, death.

The Farm's midwives and a numbers of families were painfully aware that the task of delivering babies was a life-and-death responsibility. Despite everyone's best efforts and intentions, in those early years, a number of babies were lost. The great sorrow those families endured was shared by the community as a whole and brought into focus the importance of compassion as we accept the responsibility of caring for each other.

Again, Western society in many ways has built walls separating people from direct experience. The old are placed in nursing homes where they often die alone instead of surrounded by loving family and friends. Excessive medical procedures thwart the course of nature by prolonging life, at times even against the will of the individual or their family.

The grief we experience in an encounter with death, either directly or when we lose a loved one, can bring into being our strongest connection with the profound, in the support we offer to family and friends. Each encounter with death produces a period of internal reflection and contemplation. While our contact with this experience

Care for the very old and the very young is a sacred responsibility we share as human beings, and makes us whole.

does not provide answers, it can cause a person to evaluate the meaning of their life and to consider their past actions and how they will spend the time they have left, the essence of the spiritual path and the true definition of a sacrament.

Care of the Elderly

As a direct action and an alternative to the prevalent trend in Western culture, very early on, members of The Farm began to bring aging relatives to live with them in the community. Their influence and impact has been extremely significant and an essential element in The Farm's chemistry.

The aging seniors and octogenarians who come to The Farm establish friendships and relationships that go beyond their immediate family. They are integrated and accepted with unconditional love by everyone in the community, their care a gift that encompasses both giving and receiving, often accompanied by a final blossoming of joy. As aging parents pass on and new ones take their place, their time in the community is always a gentle reminder that every person has something to contribute, that life is to be enjoyed right to the end.

The Sacrament of Marriage

To counterbalance the loose sexual attitudes of the '60s and '70s, The Farm's social code took a jump back toward the center, celebrating marriage as a sacred bond between two people. The responsibility of bringing children into the world and into a relationship was distinguished as a spiritual pact that had lifelong consequences. To elevate marriage into its proper position as a spiritual sacrament, the early Farm marriage ceremonies were performed immediately after Sunday Service meditation, the point at which the community would come together in its clearest frame of universal mind.

Marriage ceremonies express the heart of The Farm Community, and an opportunity to rejoice. Marriage clearly represents the shift from youth to adult, the connection of family, the initiation of lifelong consequences, transformation, the true definition of sacrament.

Life Force Energy

The Farm was created with the explicit intention to be a community founded on spiritual principles and values. With this ideal as its fundamental purpose, there followed an unwritten, unspoken understanding of the nature of the universe that is still shared by most of its members, forming the foundation of the community's core beliefs and agreements. The intent here is to identify and express these tenets not as absolutes, but as a context for interpretation, as a way to grasp the deeper meaning of The Farm.

Behind all aspects of The Farm's spiritual belief system is the recognition of life force energy, what established religions may refer to as holy spirit or God. All living beings contain this spark of energy, and death can be described as the absence of this same energy. We know, from deep within, the difference between even the dimmest spark of life force energy and the moment it has left the body. By consciously acknowledging the presence of life force energy, we can then learn to perceive more subtle aspects of its existence and its effect on everything around us.

Blending Science and Spirituality

Babies and young children are like fountains of energy. Their life force is strong, giving them the power needed to grow and develop. As we grow older, the intensity of our life force diminishes. When we become ill, our bodies are in a struggle to heal, calling upon our inner life force to overcome what is working against us, be it a virus, bacteria, cancer or a breakdown or malfunction of our own internal systems.

Life force energy is not just inside us, but all around us. There is no separation. Science confirms that the space between us is not empty, but a soup of molecules, atoms, protons and neutrons in all directions. The connection extends past our skins, through our skulls, to the very center of our mind and the core of our being, our soul. We are one.

The resonant patterns of energy are evident in nature from snowflakes to quartz rainbows. Over 100 years ago, scientists applied energy to crystals to generate harmonic frequencies. Technology manipulates harmonic frequencies as sound, light, radio waves, reaching further to create modern manipulations in the form of video, television, satellites and computers.

Science and spirituality are not separate. Everything in our universe has a connection to energy and harmonic frequencies.

Energy Beings

We are energy beings. Every one of us is both a transmitter and receiver of energy, human antennas sensitive to frequencies at every level. As receivers, we can walk into a room and know the energy, good or bad, high or low, warm or cold. We send out energy in everything we do, and it changes the world around us, especially when released in its highest form, love.

Feeling the vibrations, "the vibes," has gone from hippie jargon to common everyday usage, but in fact, our language has long articulated our ability to perceive vibrations, and there are multiple ways we communicate this.

We are happier when our lives are in harmony with each other or with nature. Our societies strive for racial and religious harmony. We like our music to be in harmony rather than discordant. All of these are references to smooth vibrations that resonate, producing pleasure, contrasting with feeling rattled, shook up or uptight...chaos.

Attention is Energy

> You give energy to anything that has your attention.
> You create your universe with your attention.
> What you put your attention on, you get more of.

These phrases describe the core of spiritual teachings adopted by The Farm Community. By being aware and conscious of how we direct our attention, we are able to shape our destinies and affect every aspect of our lives and the lives of those around us. Every personal achievement, every skill we learn, any task we accomplish or goal we achieve is the result of our attention. Intimate relationships fall apart when we put our attention elsewhere. Students fail when their attention is distracted or they are unable to focus their attention. Businesses collapse when not enough attention is paid to the bottom line, to employee satisfaction, to changing markets, to quality control. Wrongs are righted, the sick are healed, gardens will grow. Our personal sphere of existence is directly affected by the amount of attention we have to give and where we direct it.

Many aspects of our lives can be defined by how they relate to energy. Truth is solid, real, a positive force. Lies are words that are empty. They are not real, have no truth, no energy behind them. When we have doubt, we are holding our energy back, and are unable to move forward. While it is important to reflect, to consider, to analyze, when we have reached a decision, our strength comes from putting doubt behind us and investing 100 percent of our energy.

We know that energy is infinite, that it cannot be hoarded. When we put out energy, it comes back to us hundredfold. At the same time, we can run down our energy and must take the time to recharge. In the West, we call this the retreat or, even, the vacation.

Our energy can be taken from us through intimidation or anger. When we are able to observe these expressions of emotion as energy relationships, we can step back, name them and make the decision not to participate, or allow the intimidator to have power over us. By increasing our awareness on the nature of energy relationships, we can pay attention and make observations on a more subtle level, with a better understanding of each person's role and intentions, allowing us to react more clearly and purposefully rather than being reactionary and defensive.

Energy grows from:

Truth · Love · Kindness & Compassion · Sharing · Justice

Energy dissipates from:

Lies · Anger · Fear · Greed · Injustice

Chakras—
Energy Fields in the Body

A further example of East meets West can be observed in the terminology used to describe energy centers in the body. Eastern philosophy uses the term "chakras" to identify seven different energy centers, each one relating to a way energy manifests in our daily lives. Rather than regarding chakras as literal, this example of New Age 101 is simply another way of interpreting our relationship with energy.

Colloquial language and slang in the Western world makes very clear references to their existence, typically when the energy is seen to be out of balance. Because as humans we tend to be happiest when our lives are balanced or in harmony, mainstream society has developed language that identifies people who demonstrate energy centers in a state of imbalance.

At the same time, it is important to also observe that each of us has different strengths that may appear in people as an energy center that has greater dominance in their personality or character. Through conscious awareness of our strengths and weaknesses and their connection to energy centers in the body and spirit, we can build upon and exercise our strengths while taking any necessary steps to compensate and maintain our center of balance.

The Root Chakra is said to be located at the base of the spine. This chakra relates to stability and survival. In the West, when someone has this chakra in balance, we say they are very down to Earth or "grounded." When this energy is taken to an extreme or out of balance, the individual ensures their personal well-being while disregarding the feelings of others, referred to in the West as being a "tight ass."

The Sacral Chakra is described as emanating from the navel and encompassing the genitals, abdomen and the womb. Energies emanating from this chakra are said to include desire, sexuality, creativity, pleasure as well as pain. Addictions to sex and pornography could be considered examples of this energy out of balance. Giving in to desire can lead to bad decisions. Sadomasochism, experiencing pleasure through pain, is generally regarded as extreme behavior, an indication that the chakra's energy is out of balance in relation to others.

The chakra of the Solar Plexus above the navel relates to intuition, expressed in the West as a "gut feeling or instinct." Balance in this area is said to exemplify good self-esteem and an even personality not controlled by emotions. Joy emanates as the belly laugh, anger or fear as the tight stomach.

There is universal acceptance that the Heart Chakra is one of the most powerful forces in the human spirit. Love and compassion are considered as the most important of spiritual values, yet we also understand that even these admirable qualities can cause problems when out of balance. The "bleeding heart" or enabler can bring about harm in attempting to do good, injuring both themselves and the person they intend to help.

Communication is expressed through the Throat Chakra, the pathway to truth and honesty. An open throat chakra expressed through song has the power to deliver joy, uplifting the soul. We refer to repressed emotions as "not speaking out" or a "lump in our throats." A closed or tight throat chakra can physically manifest as a whine, not the source of the problem, but a symptom, the verbal indication of deeper issues.

The Brow Chakra relates to the pineal gland, also known as the Third Eye or mind center. This chakra goes beyond normal vision to include perception, thought and intuition. When in balance, our mind is clear. When this chakra is dominant, our social order calls that person a "brain," or "nerd," identifying a loss of social skills that could be brought in balance if that person were more down to Earth or "grounded."

The connection to spiritual values is said to emanate from the Crown Chakra, the source of all understanding, morals and principles. We convey these as trust, selflessness and a sense of purpose. If the Crown Chakra is weak, we are lost, without direction. When dominated by the ego or self, the out-of-balance Crown Chakra comes across as "holier than thou," when someone thinks they are superior to others because their spiritual values are more pure.

Our goal in naming the energy centers is to articulate the connection between mind, body and spirit. When we are unaware or deny the link between energy and how it manifests in our daily lives, we are left with a limited understanding, without the full set of tools we need to function and achieve our goals.

Telepathy

If we agree that everything is connected down to a subatomic level, then the natural extension is that our minds and thoughts are connected as well. We have all had experience with telepathy. Think of a friend, and the phone rings. We feel someone looking at us and turn around. When matched with harmony, we call it being "on the same wavelength," in tune with each other.

Learn to be aware of your feelings, your intuition, especially the initial sensitivity before you've had a chance to run it through second-level mind filters, that first flash. It is not about being judgmental, but rather about being aware, using all your senses and learning to trust yourself. Telepathy works best between people who are in sync and have an honest and open relationship. Telepathy breaks down when we lose trust in one another, allowing ourselves to put out anger and feelings are hurt.

Hippie Buddhists

From the earliest stirrings in San Francisco, there was an attraction to Buddhism, especially Zen Buddhism from Japan as taught by Shunryu Suzuki Roshi, a monk called to the West as a messenger for Buddhism in the 1960s. Suzuki taught that we must turn off the chattering monkey mind inside our heads in order that we may touch our collective big mind or God. The change we desire comes from within and can take place at any moment if we allow it to happen, with the openness of the Beginner's Mind. By making an effort to push mental

The practice of meditation allows us to separate ourselves from the chatter inside our head, aligning with the higher consciousness where we are all one.

chatter aside, the mind is allowed to clear, opening up the door to reflection, insights and greater understanding.

The story of Buddha mirrors the awakening of the hippie generation, both born in wealth, only to have the illusion shattered when exposed to the outside world and the realities of poverty, cultural and racial oppression, institutionalized by the powerful upon the weak. Like the Buddha, the founding members of The Farm Community renounced their station in life and undertook a vow of poverty and a life dedicated to the service of mankind. While this proved to be impractical and unsustainable for householder yogis raising families in the modern world, many of the tenets of Buddhism remained as consistent with core principles and values of The Farm Community.

Buddhism does not have the concept of evil, only ignorance and suffering. Ignorance can be dissipated through knowledge and awareness. Those who suffer can be healed through compassion. Buddhism does not demand a blind faith in the unknown or contend that its practitioners must convert all others to Buddhism. It strives for balance in all things, the middle road, reflecting inward, reaching outward. Buddhism's Eightfold Path is a simple blueprint that outlines the core principles for a spiritual life: right view, right intention, right speech, right action, right livelihood, right effort, right mindfulness and right concentration.

A Connection to Nature

It almost goes without saying that all members of The Farm share a reverence for nature. Free of human ego, pure in essence, the peace we feel when nature surrounds us is very real. The land is what holds the

Nature represents a direct connection to something larger than ourselves, an essence that is pure and devoid of ego.

community together, what makes it one, a whole larger than the sum of its parts. It is our church, our temple, our connection to something greater, both literally and spiritually. It is also our home, a place to work and play and live and die. We go back to this Earth.

Free Will

Each and every one of us has the power to chart our destiny and is responsible for the decisions we make. We are not controlled by stars, celestial beings, evil spirits or divine creators. We have free will. In the East, it's called karma; in the West, it's called luck. On the Farm, we refer to this as cause and effect. The Bible is in agreement: "As ye sow, so shall ye reap." "Do unto others as you would have them do unto you." Pretty clear.

Awareness and the Spiritual Path

One of the most important steps you can take is to affirm that you want a life with greater connection to spiritual values, a life of meaning and fulfillment. This is called becoming conscious or aware, to make a commitment to a spiritual path. New-age circles refer to this as personal growth, working to improve oneself. However, the spiritual path goes a bit further by acknowledging that you directly influence and create your universe through personal choices. When you set an intention and make your decision based upon spiritual values, solutions to the problems before you are more likely to appear. Hurdles fall to the side. Things work out for themselves. It comes back to the old hippie speak: Go with the flow. It is the natural flow of energy, unobstructed, like a river through the course of our lives.

It is important to remember that a path can give us direction, but we can also lose our way. Distractions, poor decisions, any number of factors can lead us away from our conscious awareness until we no longer remember where we are going. We do not see the signposts, the warnings, the red flags that come before us that are meant to capture our attention and remind us of our true intentions, to be the best person we can be. The good news is that we can find our way back, acknowledge our mistakes, forgive and be forgiven. It's what Christians call being reborn.

"Life is like stepping onto a boat
which is about to sail out to sea and sink."
— Suzuki Roshi

You better enjoy the trip!

In the end, we are all pilgrims on a journey, life's journey, with only our moral compass to guide us. With patience, integrity, honesty and hope on our side, we will find our way.

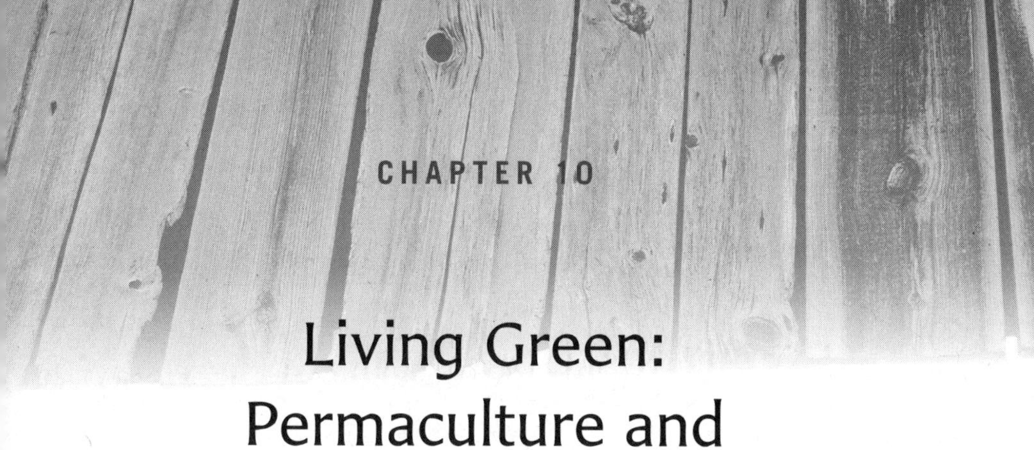

CHAPTER 10

Living Green: Permaculture and Ecovillages

Two words have come to represent the combined concepts of sustainability, human settlement and our relation to the Earth: "permaculture" and "ecovillages." More than a collection of ideas, both words have developed criteria that can be identified to measure how a community such as The Farm fulfills the standards embodied by these two terms.

Permaculture is a philosophy that takes in design and planning that includes architectural engineering, food production and structural organization modeled after natural ecosystems. Just as nature is able to continue in a revolving cycle of renewal, by emulating these patterns, new models of sustainability can be developed that emphasize the undeniable relationship between a healthy planet and healthy human existence.

Because The Farm was established many years before the term "permaculture" came into popular use, its alignment with the outlook of permaculture was not the result of conscious planning, but came from an intuitive understanding of natural order. At the same time, members of the community were among the first to adopt the permaculture school of thought, holding courses and offering internationally recognized certification in permaculture design.

Portions of the land used for long-term food production, crops such as its 5 acre blueberry orchard.

At the heart of permaculture planning and implementation is the recognition and organization of established zones. Some permaculture experts and instructors begin with Zone 0, using this designation to represent the home, its design and construction (ideally from natural materials), its energy needs and a seamless integration into the structure's surrounding environment, examining issues such as water flow and drainage. Others start with Zone 1 which takes in the area immediately around the home and includes the kitchen garden, growing crops that need daily attention. In Zones 2 ans 3, you find places for food that is grown on different timelines (plant, mulch, harvest) such as melons, sweet potatoes and winter squash, as well as field crops, orchards, ponds and hay fields, areas of the permaculture homestead that will be visited infrequently. Zone 4 takes in the semi-wild found in the neighborhoods separated by forested valleys and Zone 5, wilderness with no human intervention, such as the 1,000 acres in the Farm Community set aside as a nature preserve.

Every aspect of the zone system corresponds in some way to the Farm's use plan. Rainwater catchment from roofs, keyline plowing, ponds in pastures and examples of plant-based grey and black water filtration demonstrate not just water consumption, but a more sensible approach to water usage and methods for recharging the underground aquifer. The community's forests are a reliable and renewable source of fuel for heat, building materials and broader ecological benefits. The integration of homes along the periphery of woodlands, the mix of fields and forests, is in direct alignment with permaculture's concept of the "edge effect," where contrasting ecosystems meet. There are home gardens, community gardens, apple, pear and blueberry orchards, a community blended with nature and surrounded by land which is free from any development.

Individual homesteads, neighborhoods and entire communities can benefit from the permaculture template. In this way, The Farm makes an ideal subject that can be studied to examine ways in which it represents the design elements of permaculture. It is not that The Farm exemplifies every aspect of permaculture, but that its pragmatic approach falls in step with the philosophy's fundamental tenet, especially when looking at its broader application, a lifestyle of people in alignment with nature and as stewards of the land.

The Ecovillage Model

The word "ecovillage" implies a bond between people and nature, comprehending the unquestionable link between human settlements and their environment, going further to include the many layers of human relations. If anything, it is the disconnect from the natural world inherent in urban development that perpetuates the illusion that we are separate, isolated, and bear no responsibility to future generations. If we are to turn this spaceship Earth in a new trajectory, the inspiration will not come from corporations or governments or begin on a grand scale, but will be found in the myriad of experiments in community pointing the way. Ecovillages are like gems, tiny jewels scattered across the wider fabric of society, reflecting the light, catching our attention, a reminder that answers are often to be found in the smallest of things.

As in "ecology," the "eco" in "ecovillage" represents "ecological." The purpose of an ecovillage is to be a model of sustainable living, a way of life that is able to continue into the indefinite future because it fulfills the needs of the people who live there in a way that preserves, protects and even enhances its environment.

In the mid-90s, intentional communities from around the world gathered for a historic meeting with the objective to build a Global Ecovillage Network. The pattern of excessive consumerism and conflict driving the unbridled exploitation of the Earth's limited resources had to be addressed. For humankind to avoid its own destruction, it needs visible models, positive examples to demonstrate a holistic approach to civilization that would allow it to continue.

If the '60s communes were a search for a better way for people to live together, ecovillages are about showing the way, living examples meant to influence and alter the direction of development. In its transition from commune to intentional community to ecovillage, The Farm Community establishes its relevancy in the here and now. Although its roots were formed in '60s counterculture, the branches of its tree are living breathing examples for today reaching out into tomorrow.

Every ecovillage project around the world exhibits aspects where it shines and shortcomings to improve, and The Farm is no exception. The Farm and similar projects illustrate that there is no single solution to the world's problems, but that the road to tomorrow enters from all directions. Every community has its gifts to share, wonderful examples of the things they got right, and lessons to teach from the mistakes that were made.

It is understood that what constitutes an ecovillage has been kept broad in order to include many different types of endeavors, whether they be urban centers or rural enclaves. But if one considers the standard and accepted definition as established by leading spokespersons in the worldwide green movement, The Farm indeed fits all of the criteria of an ecovillage and then some, with many of its green attributes not immediately apparent, but embedded into the web of the community. The Farm's aim from the very beginning was to be a model of sustainability that could be emulated and adapted to serve the needs of people wherever they are, in the hope that more communities of this type would spread, flourish and multiply.

Robert Gilman, a recognized expert in sustainability and positive cultural change, has created one of the most widely accepted lists of the characteristics for an ecovillage. An overview of The Farm shows that the community stands out as an ecovillage, but that there is always room to grow.

Should be human scale

A village should be relatively small, a size that lets everyone know each other and interact on a daily basis. The network of connections becomes like an ecosystem in itself, sustaining the community.

For over 40 years, The Farm has placed a high priority on the importance of family relationships. Its population size allows members to develop multiple circles of friends and connections on many different levels. Residents live, work, play and explore unlimited ways to cooperate with each other.

The definition of an ecovillage is that it should be human scaled, permitting all residents of the community to know one another.

Social functions such as regular community dinners give Farm members a tangible way to interact, build friendships, and maintain personal connections.

Is a full-featured settlement

The Farm has many of the aspects of a small town, numerous neighborhoods, a town center, a clinic, its own water system, over five miles of roads, a dozen public buildings, a swimming hole and sandy beach, a community center, kid parks, multiple businesses and much, much more.

Farm Midwives deliver the community's newest residents. The Farm's cemetery has over 80 former (or one could say permanent) residents. Life on The Farm goes full circle.

With its own store supplying staples and a wide assortment of items, a good number of people may stay inside the community for days, even weeks on end, with no reason to make a trip into town or ever leave.

The Farm School has always been an important institution in the community. Its survival past the economic change has allowed it to

Ecovillages are also defined as full-featured settlements, supplying as much as possible, the many different and necessary components of community, such as places to work, to play, and to educate the children. Above, students at The Farm School.

continue as a focal point for community relations, bringing together young and old from across the community. People pay for student tuitions, participate as teachers, pull off fundraisers and help organize celebrations and other activities. The School also serves as a way to attract like-minded families with young children, new members bringing fresh energy to the community. Frequently families will also move into the local area specifically so their child or children can attend The School. In that way, The School extends the sense of community well beyond The Farm's borders.

One of the most important ways The Farm Community benefits its members and residents is as the hub for social activities and events. The combination of a rural lifestyle and an active social environment can fulfill the need for personal interaction, something that can be missing for those living on isolated rural homesteads. Even someone committed to city life may find The Farm offers greater opportunities for personal connection, builds deeper friendships and has a wider range of networking opportunities, while avoiding the downsides like endless traffic, a higher cost of living and other demands that eat away at personal time.

If anything, The Farm's bountiful social calendar can be almost overwhelming, with seemingly endless choices ranging from the practical, through participation on committees and governing bodies, to the ever-frequent parties, potlucks and festivities. Of course, the social dynamic can be also limited, especially for young single people who want the action of a city, and the chance to meet new people.

Ideally, an ecovillage is energy-independent, generating its own clean renewable energy. Several very large and small photovoltaic arrays produce electricity throughout The Farm, but they generate only a percentage of the power consumed by the entire community. Virtually all homes and businesses on The Farm are connected to the grid and standard electrical power. Equipment and installation costs for solar electric systems remain prohibitively high, and cannot be economically justified by the majority of families and businesses who face the same financial struggles as people do anywhere.

The solar arrays on The Farm utilize what is known as a grid tie system, in which the power they produce is fed directly back into the grid. This eliminates the need for batteries that require maintenance, have a limited life span and usually contain lead or lithium. There are tradeoffs even in green energy production that must be factored in. The Farm still holds clean and independent energy production as a goal and ideal, something that will be achieved when it becomes truly affordable.

Local food production is typically considered a fundamental aspect of ecovillages. The Farm has a large concentration of home gardens, but they supply only a portion of the food consumed by the community's residents. However, it is located in the midst of a strong agricultural area with a large population of Amish and Mennonite farmers who generate an abundant supply of local produce. Again, people on The Farm must maintain the balance between idealism and practical application. They are not separate from the realities of life in the modern world, facing the same cost of living as their neighbors in Tennessee and all Americans. Having a job or developing a career can leave little opportunity for the extensive amount of time and commitment required to come close to growing all the food needed for a complete and healthy diet. It can be far more practical for some members to purchase homegrown produce from Amish neighbors at a fraction of the cost required to grow it. They are still supporting local community agriculture, just not within the borders of the Farm ecovillage.

Gardeners on The Farm find what works best for them, whether it is fresh organic lettuce for salads, tomatoes for canning or greens for winter. The beauty of The Farm's climate is that a dedicated gardener can indeed grow nearly all of their food, a wide variety of fruits, vegetables, even grains and beans. A garden and a greenhouse can produce food all year-round. For most members, the art of gardening on The Farm is undertaken as much for the spiritual connection to the earth and the seasons, a quiet time of mediation, as the songs of birds fill the air. Gardening brings a sense of pleasure, a benefit of rural living with fruitful and delicious rewards.

At the same time, members of The Farm see farming and food production, the development of the community's orchards and management of the land as critical components of preparing for the future. Organic treatments to the fields, solar irrigation to fenced-in plots producing food and expansions to gardens are all underway. Experiments in growing protein and grains on a small scale have proven that, should the necessity arise, sustainable food production can be achieved. Beans and corn were staples in the South for generations, including for the original Native Americans, and that day could come again. So much will change over the next several decades, with many unknowns surrounding the economy, food costs and availability and climate change. The Farm's maintained fields are an investment in its future, land that can pump out a lot of food.

The Farm's internal economy also demonstrates the combined effect of collectivity and working together. Imagine the following series of events: A person hired a Farm carpenter to do repairs. The carpenter got a massage from a body worker. The massage therapist paid a friend to clean his home. The housecleaner also taught at The Farm School, but was unable to make one of her classes, so she hired yet another friend to substitute. That substitute wasn't relying on the income from her day as a teacher at The School, so she took the money and purchased art supplies for the kids. It is a win-win-win situation and a classic example of circulating money within The Farm economy before sending it out the door.

Harmlessly integrates human activities into the natural world

More than 1,000 acres of The Farm's over 1,750 acres are kept completely wild, with no development, used primarily by residents for

The Kids To The Country program embodies the Ecovillage definition of healthy human development, using the community's resources to give children from city-based homeless shelters and refugee centers the opportunity to explore the wonders of nature.

hiking and horseback riding. All residential areas are merged with the hardwood forests along ridge tops, leaving valleys separating neighborhoods undisturbed. Swan Conservation Trust, a non-profit founded by Farm members in 1994, now owns and manages an additional 1,400 acres as a nature preserve surrounding The Farm Community. Members, former members and friends of The Farm have purchased over 1,000 acres directly adjacent to the community's borders. Altogether, The Farm and its extended network currently control over 4,000 contiguous acres.

With hunting prohibited and relatively little human activity, The Farm's forests and meadows attract all sorts of wildlife, from native species to migrating songbirds. Herds of deer roam at will, to the point that all gardens must be fenced. Wild edible mushrooms sprout spontaneously and in abundance following the summer rains. Common and rare wildflowers bestow their beauty to several acres of restored prairie barrens. The Farm exemplifies a seamless integration with nature.

With so much of Tennessee's countryside either under development or converted to pasture, there are few natural meadows that are the home for certain species of birds. In recent decades, these species have been shrinking in number very drastically. In 2006, it was discovered that a rare bird, Henslow's Sparrow, had begun to nest in The Farm's fields. When first observed, it could be found at only one other location in the state. Bird-watchers from around the region came to view this small brown bird flittering about in the middle of the open spaces. One Farm member began to catalog nesting sites and monitored the sparrow population each year. The good news is that, since it was first discovered nesting in the community, Henslow's Sparrow has been spotted at other sites around Tennessee. For the time being, its place is secure on The Farm.

The Farm's extensive land base, clearly a symbol of unity for the community, relates to one of the ways it also falls outside the lines of archetypal ecovillage design. When The Farm's founders drove their buses onto the land, one of the big motivating factors was to live close to nature, as far away as possible from cities and urbanization. The lay of the land and, in many ways, the location of the logging roads running down the ridges determined the positioning of house sites and neighborhoods.

Classic ecovillage design situates housing and shared facilities in tight clusters in order to leave as much of the land free of development as possible. Ideally, all structures and other community assets are within easy walking distance, encouraging daily interaction. In contrast, The Farm's residential areas and public buildings are spread out over several miles. The School and community center are over two miles from The Farm Store and the primary residential area, making the community very dependent on vehicles and transportation. The reliance on gas-powered cars and trucks is offset somewhat by bicycles and electric vehicles, namely, golf carts, with dozens of them buzzing up and down Farm roads. Most are charged by a connection to the grid, but charging with solar power is also common at homes and workplaces throughout the community.

As The Farm looks forward at future growth and development, it must contend with the fact that new people joining the community are also seeking an alternative to urban density and a more personal connection to nature. They typically want neighborhoods with house sites that provide some level of isolation. Ultimately, this dilemma serves to demonstrate that people do prefer to have their space.

Supports healthy human development

From the very beginning, healthy human development has been a cornerstone of The Farm's mission, both inside the community and in its outreach to the wider world. The work of Plenty International, The Farm Midwives, The School and the Ecovillage Training Center all represent this commitment.

In operation for over 30 years, the Kids To The Country program brings children from homeless and refugee shelters to experience life on The Farm. Their stay includes swimming, horseback riding, bicycling, arts and crafts, as well as exercises to build skills in cooperation and conflict resolution. For many participants, it is their first time out of the city and in face-to-face contact with nature. Some kids have been able to return year after year until they eventually come back as counselors, bringing the same life-changing experiences to a new generation.

One of the reasons The Farm works is that it fits into all stages of human life. Babies are brought into the world with the tender loving care of the Midwives. Children flourish surrounded by nature and love. Young people are encouraged to express their ideals and their sense of individuality. Adults are able to exercise their vision. New families create the warmth. The grandparents, aunts and uncles of The Farm's extended family, "cresting over the hill" can slow down and take time to enjoy life. Our octogenarians are loved and appreciated, a reminder of how lucky we are to be alive.

Has multiple centers of initiative

As described by Diana Christian, author of *Creating a Life Together*, ecovillage members should have direct autonomy over the initiation

Ecovillages must provide multiple centers of initiative, empowering their residents to work together for the greater good of the community, while still maintaining individuality and personal freedom.

of their home, business, finances and other issues related to their personal life with the community.

The Farm's restructuring in 1983 was in direct response to these unsatisfied needs that were being suppressed by the original communal organization. That change put people in charge of their finances, how they earned a living, the type of home they lived in and what they spent their money on. Personal initiative on every level was the name of the game.

Since that time, Farm members have always had many ways to be directly involved in defining their own destiny and, more specifically, the direction of the community. Even relatively new members can attain positions of authority and responsibility, filling a seat on the community's Board of Directors or Membership Committee.

The community is a mix of collectivity and intense personal freedom. As long as a member maintains their minimal commitment to the community in the form of membership dues, they are free to pursue a demanding career or live a relatively low-key subsistence lifestyle. From the very beginning, the community understood that building agreements was much stronger than imposing rules. It is up to each individual to determine what motivates them and makes them happy. For the most part, anyone with a good idea who is willing to work to make it happen will find a broad base of community support. This may appear in the form of volunteers offering help and time, along with financial contributions, or at the very least, a pat on the back and a very hearty "Way to go!"

Can be successfully continued into the indefinite future

The Farm has come to understand that sustainability is very much about passing on your ideals to subsequent generations. In recent years, the community has turned a corner as new young families build homes and establish a life there, carrying forward the vision started by the founding members.

Perhaps more than anything else, the community land trust document established in 2012 ensures that The Farm's land base will remain whole and undivided, unencumbered by debt, to be a home for visionaries and dreamers for generations to come. Developed over ten years, the new land trust document was approved in a unanimous vote by the community. Its sole purpose is to protect the integrity of the land and shield it from attack or any attempt to dissolve and divide the property. This document also clarifies that its purpose is to support the beneficiaries, the members of the community.

Ultimately, the future of The Farm comes not from a legal document filed at the courthouse, but in safeguarding the values that make it a place where people want to live. As the great burst of energy, the spark of selflessness that led its founders to this land recedes farther and farther into the distance, it will be up to future members, the next generation and the one after that, to remember that it is not bylaws, or money or the land itself that will hold the community together into the distant future. As it was in the beginning, it will be the personal bonds, forged over time, through shared common experience, that will maintain The Farm as a symbol of hope and freedom, a beacon of inspiration and a model for how we were meant to live.

CHAPTER 11

The Future Farm

THE FOUNDING OF THE FARM was a direct response to what might be called the signs of the times. Embroiled in a seemingly endless cycle of war, the front line of capitalist imperialism and neocolonialism could not be ignored, expressed daily in the jungles of Vietnam. Corporate interests were aggressively plundering the world's dwindling resources, leaving an environment devastated by pollution in its wake. Unbridled consumerism sacrificed quality of life for quantity, marching headlong toward a precipice clearly visible to eyes open and aware that everything is connected.

The assassinations of Martin Luther King, Jr., the voice exposing racism and a messenger of peace, and Robert Kennedy, the antiwar candidate; the students at Kent State University, murdered by the National Guard; followed by the landslide presidential election of Richard Nixon—all these events and many more made it painfully obvious that change would not come from within the system. For the group landing in Tennessee, the revolution was in a new direction, to build an ark that could ride out the inevitable storms of an uncertain future.

The world of the new millennium is in no better condition. War continues to dominate and affect every aspect of our lives, fueled by the forewarned military industrial complex, oil interests and ideological conflict. Young people continue to sacrifice their lives and limbs, swept up in patriotic rhetoric, their sense of self wracked by an ethos of violence, left abandoned to cope with guilt and depression. A treadmill of fear plays into the hands of increased surveillance and the sacrifice of freedom. Oil spills, nuclear meltdowns, devastating storms, hotter temperatures driving droughts and a season of fire, mountaintop removal and poisonous fracking, rainforests laid to waste at an increasingly rapid pace—it's obvious little has been learned from mistakes of the past and that caring for the Earth is in no way a priority of the government or corporate agenda.

The course charted by The Farm is one that continues to be a viable alternative lifestyle and an achievable route to positive personal fulfillment. At the center lay key concepts and definitive actions that are as relevant today as they were when The Farm was first created.

Use nature as a focal point, a means to bring peace and a connection to something greater than yourself directly into your consciousness and existence. Pursue right livelihood and develop a way to support yourself that brings satisfaction, makes you feel whole and has a positive impact on the world. Consider acquiring land, the base for a home that is not controlled by a bank, moving you one step closer to independence, a buffer from the stresses that can distract us from that which is truly important. Consider acquiring land, as a way to grow food, to know its source, to touch the earth, a gentle reminder of life's changes and cycles. Make a deliberate effort to find community and use it as a way to both receive and offer support to a wider network of family and friends.

Community does not have all the answers, but it does speak to the questions that linger in our hearts and minds: Am I satisfied with the path that I have chosen? Does it lead to sanity and contentment? Will it provide a safe and peaceful existence for me and my family? Am I taking steps that will carry me toward my true aspirations? What can I do to bring about change in the world? Where will I go from here?

Philosophies of the East describe a systematic plan based on the wisdom of the farmer. You begin by tilling the square inch, move out to the square foot and on to the square yard. In other words, only by working on ourselves, starting from within, the square inch, can we sweep away the cobwebs of doubt in order to move forward with clarity and a sense of purpose. Nurture the family, care for it as you would the fruits of tomorrow's harvest, the square foot. Grow community, wherever you are, seeking out others with a common vision, the square yard. Remember, we are stronger together than any one of us alone. We are the lever and community is the fulcrum, amplifying our efforts beyond what we can accomplish as individuals.

As Margaret Mead stated so elegantly, "Never doubt that a small group of thoughtful, committed people can change the world. Indeed, it is the only thing that ever has."

The Future Farm is not on a piece of land in Tennessee. It lives in the heart of every person who has visions of a green, sustainable, peaceful planet. Those dreams come to life when we turn ideas into action, come off the sidelines and become active players in the game.

Look into what The Farm calls "your heart of hearts," your deepest, most sincere feelings, and determine your ultimate goals, where you want to be 20 years down the road as your objective. Knowing your destination, where will you be at halfway, ten years from now? In five years you could have a degree, a career, a down payment, a reachable achievement. In order to get there, where will you be in two years? One year from now? What will you do today?

The Farm exemplifies the definition of "synergy," from the Greek "synergos," which means "working together," a living demonstration that the energy of the whole really is greater than all of its parts. Everything it has accomplished began with an individual setting an intention, someone who made a decision that it was better to try and fail than never to have tried at all. The Farm as an entity exists because enough people understood from the very beginning that only by relying on each other can we truly succeed.

Experiments like The Farm will continue to emerge throughout the course of time because every person knows somewhere deep inside that humanity's struggles are the result of choices, and we can do

The solutions will come from the small circles of people who choose to live, to act, to be, in community.

better. There will always be people who are called to come together, to reignite the vision that we are one people on one planet in one Universe. The solutions will come from the small circles of people who choose to live, to act, to be, in community.

Index

COVER IMAGE CREDITS: Upper left—Buses, photo by Gerald Wheeler, courtesy The Farm Archive Library; Top row, center—Douglas on the ham radio, courtesy Plenty International; Top row, left—Group, courtesy Douglas Stevenson; Middle row, left—Douglas family, photo by Clifford Chappell, courtesy The Foundation, The Farm Archives Library; Middle row, right—Douglas, VW van, courtesy Douglas Stevenson; Middle row, right—Tractor, courtesy, The Foundation, The Farm Archives Library, previously published in *Hey Beatnik*, 1974; Middle row, color, left—Teacher and kids, photo by Douglas Stevenson; Middle row, color, right—Girl with headset, photo by Douglas Stevenson; Middle row, left—solar panels, photo by Douglas Stevenson; Middle row, center—Group of men, photo by Anita Whipple; Middle row, right—Steel dome with white cover, photo by Douglas Stevenson; Bottom row, left—Yoga, photo by Douglas Stevenson; Bottom row, center—Midwife, mom and baby, courtesy The Foundation, The Farm Archives Library; Bottom row, right—The Solar School, photo by Douglas Stevenson.

BOOK IMAGE CREDITS: All photos in the book are by Douglas Stevenson, except those noted here: Page 2, Gerald Wheeler; page 7, previously published in *Hey Beatnik*, 1974, photographer Daniel Luna; page 68, Plenty International; page 74, Plenty International; page 86, previously published in *Hey Beatnik*, 1974; page 90, previously published in *Hey Beatnik*, 1974, photographer Gerald Wheeler; page 140, previously published in *Hey Beatnik*, 1974, photographer Gerald Wheeler; page 141, previously published in *Hey Beatnik*, 1974; page 148, previously published in *Hey Beatnik*, 1974, photographer Douglas Stevenson; page 157, previously published in *Hey Beatnik*, 1974; page 168, Plenty International, photographer Don Edkins; page 175, Plenty International, photographer Dennis Martin; page 178, Plenty International; page 188, previously published in *Hey Beatnik*, 1974; page 213, Anita Whipple.

About the Author

DOUGLAS STEVENSON has been a member of The Farm Community for 40 years. He has been a volunteer with Plenty International, the community's relief and development nonprofit, and is an active board member of Swan Conservation Trust. Along his journey with The Farm, Douglas has served on the membership committee and on the board of directors, as well as spending eight years as its manager. He is the de facto public relations person, and has helped present The Farm's story to countless newspapers, magazines, documentary film makers and television journalists. His company Green Life Retreats hosts the Farm Experience Weekend and other instructional seminars about sustainable living (www.greenliferetreats.com). Douglas is also available as a public speaker. For more information visit douglasstevenson.com.

If you have enjoyed *The Farm Then and Now* you might also enjoy other

BOOKS TO BUILD A NEW SOCIETY

Our books provide positive solutions for people who want to make a difference. We specialize in:

Sustainable Living • Green Building • Peak Oil • Renewable Energy • Environment & Economy Natural Building & Appropriate Technology • Progressive Leadership Resistance and Community • Educational & Parenting Resources

New Society Publishers

ENVIRONMENTAL BENEFITS STATEMENT

New Society Publishers has chosen to produce this book on recycled paper made with **100% post consumer waste,** processed chlorine free, and old growth free.

For every 5,000 books printed, New Society saves the following resources:[1]

25	Trees
2,264	Pounds of Solid Waste
2,491	Gallons of Water
3,249	Kilowatt Hours of Electricity
4,115	Pounds of Greenhouse Gases
18	Pounds of HAPs, VOCs, and AOX Combined
6	Cubic Yards of Landfill Space

[1]Environmental benefits are calculated based on research done by the Environmental Defense Fund and other members of the Paper Task Force who study the environmental impacts of the paper industry.

For a full list of NSP's titles, please call 1-800-567-6772 *or visit our website* at:

www.newsociety.com